PROBABILITY THEORY, LIVE!

PROBABILITY THEORY, LIVE!

More than Gambling and Lottery—It's about Life!

Ion Saliu

Library of Congress Control Number: 2010901441
ISBN: Hardcover 978-1-4500-3735-8
Softcover 978-1-4500-3734-1
Ebook 978-1-4500-3736-5

This book was printed in the United States of America.

To order additional copies of this book, contact:
Xlibris Corporation
1-888-795-4274
www.Xlibris.com
Orders@Xlibris.com
73470

CONTENTS

Chapter I

THE FUNDAMENTALS OF PROBABILITY THEORY

1. Necessarily the Best Introduction to Theory of Probability

The most important questions of life are, for the most part, really only problems of probability.

—Pierre Simon de Laplace,
Théorie analytique des probabilités

I view a best introduction as necessarily the best treatise. A good thing is better when it has a good beginning. Science goes sometimes into such details that make it sound like a jargon-speak. Theory of probability has, undoubtedly, its jargon. It also has that huge number of formulas and equations with no practical purpose but to torment sleepless students before exams.

I get my share of questions regarding various aspects of probability. I can also see in public forums plenty of probability problems. Who can answer all those questions? What I remark, however, is *a deficiency in the introduction to theory of probability*. Probably the essentials are skipped too fast in order to cover all those insomnia-causing topics.

I wrote previously a few Web pages dedicated to probability and odds (at SALIU.COM). Since I am unable to respond to most private questions and requests, I try to put together now the essential introduction to theory of probability. We must start with the start: the mother of all probability formulas, the formula that gives birth to many other formulas. We must also formulate fundamental algorithms of analyzing a wide variety of probability problems. Then we must put on the table the most efficient instrument of answering (almost) all questions on probability.

Furthermore, contemporary theories, especially in mathematics and sciences, greatly benefit from computers and software. I am one of the few fortunate ones: I am also a computer programmer. I've been able to comprehend a multitude of probability problems and write adequate software to solve such problems—and then some.

Historically, by a majority of the accounts, theory of probability started in seventeenth-century France. The father of this branch of mathematics was ***Blaise Pascal***, a writer, a philosopher, a mathematician. He was also a theologian who considered a *wager* as to the existence or the inexistence of God! I would win huge against you, Blaise, *mon ami*! I know the formula that you didn't! *Vive l'ordinateur!*

Blaise Pascal was posed with two famous problems by a noble named Chevalier de Méré. M. de Méré was a gambler, especially in games of rolling the dice. In the first problem, de Méré was confused about the odds of getting at least one six-point face rolling one die four times. The problem looked very complicated at that time. Now we can easily do such calculations by applying the Binomial Distribution Formula

(*BDF*, read on in this chapter). BDF calculates the probability of *exactly* M *successes in* N *trials*. I created software to expand BDF to all three cases: *at least, exactly, at most.*

The other de Méré problem (or case) was even more difficult and controversial. Similar cases still produce controversy and confrontation today. It is about the role of the *past* in random events. The dominant opinion is that past events have absolutely no correlation to future events. They say, for example, that history (past lottery drawings) has no influence over the next drawings.

Two gentlemen played a best-of-three backgammon match. The game had to be interrupted after the first game. Thus, one of the gentlemen had a 1–0 lead. The two gentlemen agreed to resume the game as soon as the circumstances permitted. The gentleman in the lead wanted to continue with game 2. The competitor who was trailing wanted a fresh start. Restart at 0–0, that is.

Blaise Pascal demonstrated that the player in the lead had already an advantage he *had earned*. Pascal was new to the field of mathematics known as theory of probability. As a matter of fact, Pascal was inventing probability theory! So his calculations looked rather complicated, nonetheless convincing.

The past counts. We must suppose that the two players are equally skilled; thus, the winning probability is 1/2 or 0.5. The first player to win 2 games wins the match. The trailing player must win the next 2 of 2 games. The binomial distribution probability for 2 successes in 2 trials is 0.25 or 1 in 4. The player with the lead needs to win only 1 of 2 games. The binomial distribution probability for at least 1 success in 2 trials is 0.75 or 3 in 4. Clearly, the player in the lead has a better chance to win the match after its resumption.

2. The Logical Foundation of Probability Theory

Probability starts with logic. There is a set of *N* elements. We can define a subset of *n* favorable elements, where *n* is less than or equal to *N*. Probability is defined as the rapport of the favorable cases over total cases, or calculated as follows:

$$p = \frac{\mathbf{n}}{\mathbf{N}}$$

It is the ***Fundamental Formula of Probability (FFPr)***, and everything in theory of probability is derived from it.

A simple example of sets and favorable elements: There are 10 balls in a jar; 5 of the 10 balls are red. What is the probability to extract one red ball from the jar? There are 5 favorable cases in a total of 10 elements; the probability is 5/10 or 1/2 or .5.

The easiest case is coin tossing. What is the probability of heads in 1 coin toss? Answer: *1/2* or *.5*.

The probability can be also understood as expected number of successes in one trial. That formulation makes it easier to understand why probability can never be higher than 1: no event can have more than one success in one try! The flip side is that the phenomena with only one side do not exist. TheEverything does have at least two sides. In fact, TheEverything is a unity of two opposites—the all-encompassing Flipping Coin paradigm.

The ***odds*** are often used in theory of probability, especially the branch dealing with gambling. The most correct usage of odds points to the degree of difficulty for an event to occur. The odds are *N* to *n* or (*N* – *n*) to *n*. It is widely used in horse racing. *The odds for Horse X to win are 5 to 2.* The ***N***-to-***n*** case represents the *odds against*.

Another form of odds expression is used now even more widely than the term *probability* or *probabilities*. It is the so-called *favorable odds*. The *favorable odds* are calculated and expressed as ***n in N***, most frequently as ***1 in N1***. If probability is most often expressed as a floating point number between 0 and 1, the favorable odds tend to be expressed as 1 in *Number_Greater_Than_1*, usually an integer. *The odds of winning exactly 3 of 6 in a 6/49 lotto game are 1 in 57.* That *odds* term doesn't sound grammatically correct because of the *1-in-M* form of expression. Personally, I try to always use *probability* instead of *odds*. In fact, it is the absolutely correct form of expression. The odds came from horse racing and the old-time Brits who loved that gambling activity.

I prefer expressing the probability as two terms: ***n/N*** and ***1 in N1***. I never use the expression *N to n*. Let the old-timers use that when they bet on the nags.

3. Calculating the Essential Elements of Probability

We know a lot about probability if we can determine these two fundamental elements:

~ number of favorable cases, ***n***
~ total possible cases, ***N***

In many cases, the total number of possible cases is calculated by *combinatorics*, a branch of mathematics. Still at my Web site, you'll find the best *presentation of combinatorics* (ease of understanding). All four numerical sets are clearly presented. Moreover, free software is provided to generate all possible types of sets and also calculate total possible elements in the sets—*PermuteCombine.EXE*.

In other cases, calculating total number of cases is a matter of *enumeration*. How many sides does the coin have? Two, that's the number of total cases. A die has six faces—that's total possible cases.

Determining the number of favorable cases is more difficult in many other cases. In a very simple case like betting on heads in coin tossing, the determination of *n* is easy. There is one favorable case out of two. Betting on face 6 of a die is also easy: one favorable case (out of six).

The lotto games present an even harder case of calculating ***n*** and ***N***. Total possible cases, ***N***, can be calculated by a well-known formula: *Hypergeometric distribution probability*. Or we can apply algorithms of *lexicographical indexing*, another chapter of combinatorics. Then we need to enumerate or use software to count all favorable occurrences (cases) of ***n*** and ***N***.

4. Precise Definition of the Probability Events

If we define precisely the probabilistic events, we can come up with a template of resolving just about every probability problem. I believe that the following two pairs of attributes are the essential foundation of the ***probability template***:

~ separable–inseparable
~ single trial–multiple trials

4.1. Separable-Inseparable Events

Is the coin separable in two sides? No! Therefore coin flipping is an inseparable event. Is the die separable in six faces? No! Therefore dice rolling is an *inseparable* event.

How about the lottery? Well, there are a number of balls inside a drawing machine. A number of balls (e.g., 6) are separated from the total number of balls (e.g., 49). Therefore, a lottery drawing is a *separable* probability event.

Based on this attribute, we can calculate the probability differently. In general, in the first case, we apply the very first formula of probability ($FFPr = n/N$) or the probability of the normal distribution (a more general and encompassing case than the first formula). In the separable case of probabilistic events, we calculate the probability by applying the hypergeometric distribution.

4.2. Single Trial–Multiple Trials

If we toss the coin once, that's a *single-trial* probability event. We roll the die once, also a single trial. The probability is calculated more easily in such situations. In general, the n/N formula is applied. One in two (1/2 or .5) is the probability to get heads in one coin toss.

The questions can get more and more complicated. That's how the so-called probability problems come to life: *multiple trials*. What is the probability to get heads 5 times in a row? What is the probability to roll face 6 of a die exactly 4 times in 10 throws? How about the probability to get at most 4 heads in 15 coin tosses? Or what is the probability of getting at least 4 winners out of 6 numbers drawn in a 6/49 lotto game? To answer such questions, we need to apply more sophisticated and complicated formulas and algorithms and even computer programs.

5. The Probability of Binomial Distribution: Inseparable Events, Multiple Trials

The formula is also known as the *probability of repeated trials*. What is the probability of tossing exactly 5 heads in 10 coin tosses? M *successes in* N *trials* is yet another definition for this type of probability problems. The formula relies on factorials (!) and combinations functions (*C*).

$$\mathbf{BDF} = \frac{N!}{(N-M)! * M!} * p^{M} * (1 - p)^{N-M}$$

Or using a simpler equation of combinations, *C (N, M)*, as follows:

$$\mathbf{BDF = C\ (N, M) * p^{M} * (1 - p)^{N-M}}$$

BDF = the binomial distribution probability
p = the individual probability of the phenomenon (e.g., p = .5 to get tails in coin tossing)
M = the exact number of successes (e.g., exactly 5 tails in 10 coin tosses)
N = the number of trials (e.g., exactly 5 tails in 10 coin tosses = number of trials).

6. The Hypergeometric Distribution: Probability Separable Events, Single Trial

The formula is also known as the probability of exactly *M* successes of *K* elements drawn in a pool of *S* favorable elements from a total of *N* elements. Let's take as an example a 6/49 game. The lottery commission draws exactly 6 winning numbers. The player must play 6 numbers per ticket, but the player has the latitude to select from a preferred pool of numbers (e.g., 12 numbers with good frequency). The question becomes what is the probability of getting exactly 5 winners out of 6 in my pool of 12 numbers from a total of 49 lotto numbers. The formula of probability of hypergeometric distribution answers a question that sounds very complicated, but mostly because of wording.

$$\mathbf{P(m \text{ of } k \text{ in } s \text{ from } n)} = \frac{\mathbf{C(n, k)}}{\mathbf{C(s, m) * C(n - s, k - m)}}$$

The *hypergeometric distribution probability* formula has certain restrictions. They are nasty, especially for a computer programmer trying to implement probability algorithms. Some cases are logically impossible (e.g., 1 of 6 in 10 from 10).

7. The Probability of Combined Events, or More Complicated Problems, Etc.

I don't think there are probability problems without an answer. The formulas and algorithms above are the necessary and sufficient instruments. It's all a matter of logic. Determine total possible number of cases, determine the number of favorable cases, try first the fundamental formula of probability (*n/N*). If it doesn't work, use either the binomial distribution or hypergeometric distribution. If still the answer is not convincing, use the results in deeper algorithms.

A problem like this one shows up in forums and newsgroups and e-mails. Let's throw four dice. What is the probability to get all faces equal (e.g., 1-1-1-1)? This is simple actually. No hypergeometric or binomial is necessary. Total number of cases: 64 = 1,296.

There is a total of 6 favorable cases, from 1-1-1-1 to 6-6-6-6.
Clearly, the probability is *(n/N)*

p = 6/1296 = .004629 (1 in 216).

The case can be further complicated by asking what is the probability to get at least one pair (e.g., 1-1 or 3-3). The total number of cases is the same as above. The 6 previously favorable cases can be broken down in cases of at least 2 faces being equal. The probability for *at least* or *at most* cases can be calculated using my free software *FORMULA.EXE* and especially *SuperFormula.exe*. First, we can calculate the probability for each of the 6 faces. To get at least 2 of the same face throwing 4 dice, there are 6 possibilities. Since we have 6 pair possibilities, the number of favorable cases becomes 6 * 6 = 36. Finally, the probability to get at least 2 of the same face when throwing 4 dice is 36 / 1,296 = .2778 (1 in 36).

Wrong! It isn't a separable event, remember? It is a type of the so-called *Birthday Paradox* (or probability of duplication or collisions or repetitions).

8. The Probability of Inseparable Events, Single Trial

The calculations are based on what I call *exponential sets* or *Ion Saliu's sets*. The *Birthday Paradox* is a particular case of *exponential sets*. Such sets consist of unique elements and duplicates. The unique part of an exponent is equal to the arrangements type of the set. The part with at least two elements equal to each other is the difference between exponents and arrangements. In the probability problem of four dice, the Birthday Paradox parameters are as follows: (a) lower bound = 1, (b) upper bound = 6, and (c) total elements (number of dice) = 4. The probability to get at least two dice showing the same point face when throwing four dice is .7222 or 1 in 1.385. Easy to verify. Throw four dice. In almost three out of four rolls, at least two dice show the same face.

Also, the probability to get the four dice to show the same point face is precisely calculated by using the *exponential sets*. A die has six faces—always! The probability to get exactly 1-1-1-1 is 1/1,296. The probability to get exactly 6-6-6-6 is 1/1,296. The probability to get exactly 1-2-3-4 is 1/1,296.

The pick 3 and 4 lottery games should be considered forms of dice rolling, therefore inseparable phenomena. A drawing machine is a 10-face die. The pick 3 game is like casting three 10-faceted dice. The slot machines, by extension, are the equivalent of casting multifaceted dice (usually 3 dice).

Here's another probability problem that pops up in forums and newsgroups and e-mails. A jar contains 7 red balls, 6 black balls, 5 green balls, and 3 white balls. We can construct a huge variety of probability problems with the 21 balls. For example, the probability to draw exactly 5 balls with this exact composition: 2 red, 2 black, 1 white. We must apply the hypergeometric distribution of each color.

- exactly 2 red of 5 drawn in 7 red from a total of 21 balls: .375645 (1 in 2.662)
- exactly 2 black of 5 drawn in 6 red from a total of 21 balls: .335397 (1 in 2.982)
- exactly 1 white of 5 drawn in 3 white from a total of 21 balls: .4511278 (1 in 2.217)

The combined probability is the product of the three: .056838 or 1 in 17.6.

How about the probability to draw 5 balls and get at least one ball of each color? Applying now the option of *SuperFormula.EXE* ("Win at least Lotto and Powerball"):

- at least 1 red of 5 drawn in 7 red from a total of 21 balls: .9016 (1 in 1.109)

- at least 1 black of 5 drawn in 6 red from a total of 21 balls: .8524 (1 in 1.173)
- at least 1 green of 5 drawn in 5 green from a total of 21 balls: .7853 (1 in 1.273)
- at least 1 white of 5 drawn in 3 white from a total of 21 balls: .5789 (1 in 1.727)

The combined probability is the product of the four: .3494 or 1 in 2.86.

9. The Probability of Separable Events, Multiple Trials

The fourth category of probabilistic events can be solved by breaking down the problem into two steps. Take for example the lotto games. What is the probability to get *exactly 2 hits 4 of 6 in 10 draws*? First, apply the *hypergeometric distribution probability* formula as in the case of separable events, single trial. We get a probability, *p1*. Next, we apply the *binomial distribution probability* for *exactly* M *successes in* N *trials for an event of probability* p1. All done!

10. Nothing without a Degree of Certainty—*Ion Saliu's Paradox of* N *Trials*

As long and complicated as it may be sometimes, calculating the probability, *p*, is only the first step! If we determine that the ***probability***, ***p***, equals 1 in something, it does NOT guarantee that the event will take place in a ***number of trials***, **N**, equal to something. You flip the coin 2 times. The probability to get heads is 1 in 2. You expect 1 heads, but don't hold your breath! The ***degree of certainty***, **DC**, is 75% that heads will come up in 2 tosses. In a significant percentage (25% of cases), heads will not show up. That's far from negligible.

The degree of certainty can be viewed as a probability of probability strongly connected to a number of trials. The master formula that calculates the number of trials, ***N***, for an event of probability, ***p***, to appear with a degree of certainty, ***DC***, is known as the ***Fundamental Formula of Gambling (FFG)***. We may also call it the *Fundamental Formula of the Universe* or the *Formula of TheEverything*.

$$N = \frac{\log(1 - DC)}{\log(1 - p)}$$

If $p = 1 / N$, we can discover an interesting relation between the degree of certainty, *DC*, and the number of trials, *N*. The degree of certainty has a limit when *N* tends to infinity. Let's analyze a few particular cases.

• Rolling the unbiased dice—actually, just one die. The probability to get any one of the point faces is p = 1/6. The degree of certainty, *DC*, to get any one of point faces in 6 throws is 66.5%.

• Spinning the roulette wheel. The probability to get any one of the 38 numbers is p = 1/38. The degree of certainty, *DC*, to get any one of the numbers in 38 spins is 63.7%.

• Let's look at a case with a very large number of possibilities, therefore a very low probability—a 6/49 game. Total possible combinations in a 6/49 lotto game is 13,983,816. The probability to get any one of the combinations is p = 1/13983816. The degree of certainty, *DC*, to get any one of the numbers in 13,983,816 drawings is 63.212057% (0.63212057).

• The limit of the *degree of certainty*, DC, is **{1 - (1/e)}** when N *tends to infinity*, for an event of probability *p = 1/N* and a *number of trials equal to* N. The variable ***e*** represents the base of the natural logarithm and equals approximately *2.718281828 . . .*
The limit **{1 - (1/e)}** is approximately ***.63212055 . . .***

• Let this mathematical relation be forever known as ***Ion Saliu's paradox (problem) of* N *trials***.

• The opposite problem is an older one. It is known as the ***Problem of No Coincidences*** or ***Montmort's Problem***. It was named after the mathematician Pierre Rémond de Montmort and was published in 1708. Montmort solved the problem in a very complicated manner, by approximating. He *approximately* formulated that the probability of *no coincidences* or *no matches* was the following:

$$p = 1/2! - 1/3! + 1/4! - 1/5! + \ldots$$

This calculation is very imprecise for few elements (especially fewer than 8). The degree of precision grows dramatically with the increase in the number of elements. I discovered the most precise formula of calculating the *probability of no coincidences* as the opposite method of calculating *Ion Saliu's Paradox*. The degree of certainty is precise for any number of elements (higher than 1; N = 1 leads to an absurdity). The limit is **1/e**. The *Montmort's Paradox or Problem of No Coincidences* is detailed in chapter 7.

The general formula of probability, *p = n / N* (favorable cases over total possibilities), does not always lead to *1/{Integer}* cases (e.g., {1 in *N*}). For example, play 3 roulette numbers at a time. In this example, n = 3, and N = 38. The division 38/3 does not result in an integer. The Ion Saliu Paradox is not limited to {1/Integer} cases. In the roulette case here, playing 3 roulette numbers in 38 spins, the paradox leads to this result:

{1 – [(1/e)^3]} = {1 – [0.05]} = approx. 0.95

The generalized *Ion Saliu Paradox* for $p = n / N$ and N trials:

The degree of certainty, *DC*, tends to $\mathbf{\{1 - [(1/e)^n]\} = [1 - (e)^{-n}]}$, when N tends to infinity regardless of probability, p.

There are people, indeed with training in mathematics and probability theory, who don't accept the idea that a mathematical *limit* can be reached both from the left (*incrementally*) and from the right (*decreasingly*). The limit of this probability paradox is reached decreasingly whereas it is typically viewed as an incremental limit. Run my great program *SuperFormula.EXE*, the *C* option: Degree of Certainty, DC.

• Is it the same difference—the same thing gambling this way? For example, *play one lotto ticket* **N** *consecutive drawings* or **N** *tickets in one drawing*. Something is very clear now. If you play 1 roulette number for 38 spins, you are not guaranteed to win! You have only a 63.7% chance to win. On the other hand, if you play all 38 numbers in 1 spin (trial), you are guaranteed to win! You, who have ears to hear, don't bet it all on one spin or number. Play more numbers or tickets at once. The probability is significantly lower for two or more numbers or combinations to have simultaneously long losing streaks.

I wrote also software to simulate Ion Saliu's Paradox of N Trials: OccupancySaliuParadox.EXE, a free software.

11. The Necessary Tools to Calculate and Verify Probabilities: Software

This is such an important step. We can make lots of mistakes doing the calculations by hand or even using calculators. The more complicated the formulas, the higher the probability to make errors. The computer programs are of invaluable help, really. I know it firsthand. Moreover, good computer programs can also generate all the elements of various sets for various probability cases. It is the case above, the one involving four categories of balls. You can use *PermuteCombine.EXE*, the *Word* option for *Combinations* (type of sets). You first need to write a simple text file consisting of twenty-one lines. Each line consists of one letter (the beginning of each color) from R R R . . . to . . . W W W.

If you select the *Lexicographical* generation, there are 20,349 combinations. It would be hard to count all configurations such as RRBBW, WRBBR, etc. It would be easier to generate just 100 random combinations. The results should be close to the theoretical probabilities. I tested both ways, using also in-house software. This is *the* most accurate method of determining the real probability. Generate ALL possible elements of the set (in our case, 5-letter combinations RRBBW, WRBBR, etc.) Then count FAVORABLE occurrences (e.g., Red-White-Black-Red-Black, White-Black-Red-Red-Black, etc.)

The following are more free software that I created and you can use with no problem or strings attached:

• *SuperFormula.EXE* is the best probability software—period. Among many other functions, the program does a multitude of probability calculations: exactly, at least, and at most. SuperFormula.Exe also calculates the relations between the probability, *p*, the degree of certainty, *DC*, and the number of trials, *N*.

• *FORMULA.EXE* does the calculations for the Fundamental Formula of Gambling as presented at Gambling Formula. The program can perform plenty of probability calculations. A 16-bit software superseded by SuperFormula. EXE.

• *OddsCalc.EXE* calculates the odds of any lotto game, including Powerball and Keno. If the game draws 6 winning numbers, the program calculates the odds from 0 of 6 to 6 of 6. Of course, 6 of 6 represents the jackpot case.

• *Odds.EXE* calculates the lotto odds using the *hypergeometric distribution probability*. The odds are calculated as **k** *of* **m** *in* **n** *from* **N**. More clearly, let's suppose a lotto 6/49 game. The lottery draws 6 winning numbers. The player must play exactly 6 numbers per ticket. But the player can choose to play a pool of 10 favorite numbers. What is the probability to get 4 of 6 in 10 from 49? The favorable odds: 1 in 90.

• *PermuteCombine.exe* is the universal permutations, arrangements, and combinations calculator and generator for any numbers and words.

• *LexicographicSets.EXE* is the universal permutations, arrangements, and combinations lexicographic indexing (ranking).

• *OccupancySaliuParadox.EXE* simulates Ion Saliu's Paradox of *N* Trials (Drawings).

• *BirthdayParadox.EXE* is based on the popular probability problem known as the *Birthday Paradox*. It is well presented by Warren Weaver in his famous book ***Lady Luck*** (p.132). It is a great book for popularizing complex matters such as theory of probability.

Lady Luck inspired me a great deal. I have stepped forward, however. My main advantage is the ability to write software and validate or invalidate many rules in probability theory (including *paradoxes* and *caveats*).

• *Collisions.EXE*: The *Birthday Paradox* is one tiny particular case derived from the mathematical sets named *exponents* or Saliusian sets. The *Ion Saliu sets* are the best tools to calculate a wide variety of probability problems. *Collisions.EXE* deals specifically with the probability of *collisions or duplication/duplicates or coincidences or repetition*. What is the probability of generating *N* random numbers with at least TWO of the numbers being exactly the same (duplicates)?

• *BirthdayParadox.EXE* works best with birthday cases (i.e., smaller numbers 1 to 365). *Collisions.EXE* works best with larger numbers, such as genetic code sequences, lotto combinations, social security numbers, etc. *Collisions.EXE* is less accurate with small numbers, such as birthday cases (e.g., inaccurate for birthdays of 200-plus persons in the room).

All software listed here—and much, much more—are available from my Web site, SALIU.COM. Downloading is available to registered members only. The membership is permanent in exchange for a reasonable fee. The software itself is ***free*** to use for an unlimited time. There are over one hundred software titles available at the time of this writing.

The applications of probability theory to specific phenomena will be presented in dedicated chapters. Specific software will be also presented in the corresponding chapters.

We will analyze the application of theory of probability to such phenomena as gambling and lottery. The emphasis will be on the two types of phenomena. This book will not be, however, a manual like the discredited *How to Win the Lottery* or *How to Win at Roulette*!

More phenomena will be thoroughly analyzed as well. How about the chance of at least two identical DNA sequences or social security numbers? How about the possibility of intelligent life in at least one other place in this Universe?

Reading a book in the Internet era is a much more enriching experience. Instead of footnotes or notes at the end of the book, we just do a Web search. A whole lot more is revealed—and we can use it in conjunction with the book we read. We are witnessing the beginning of the end for the printed book. So many printed newspapers died in 2009! The coldhearted truth is that reading a publication online (e-book or Web page) is far more useful than in print. My computer connects me to the entire world. Virtually all human knowledge is available to me!

So many resources are available online as complements to a book: multimedia resources, opposing viewpoints, and yes, software!

Chapter II

COMBINATORICS: THE MATHEMATICS OF NUMERIC SETS

---

1. Introduction to the Mathematical Sets of Numbers: Exponents (Saliusian Sets), Permutations, Arrangements, Combinations

We saw in the introductory chapter that *probability theory* is founded on *logic*. There is a set of N elements. We can define a subset of n favorable elements, where n is less than or equal to N. Probability is defined as the rapport of the favorable cases, n, over total cases, N.

It is obvious now that probability always deals with *discrete* elements (i.e., elements we can always count). And thus our main task is to take inventory of all elements in the set and then determine the group of favorable cases.

In most cases, such a task is very easy. We can easily count the faces of a die: *6*. Next, we only have to state what probability we want to calculate. Let's take the easiest example: the probability to get any point-face (e.g., 5) in 1 roll of the die. There is exactly one (1) point-face reading *5*. Therefore, the probability to get a *5* in one roll of the die is *1/6* or *1 in 6* or *.166666666 . . .*

That easy situation of enumeration is quite rare in the grand scheme of things. In many cases, the total number of possible cases can only be calculated by *combinatorics*, a branch of mathematics.

Any finite number of elements can be put together in groups based on certain rules. Such groups are known as sets. The sets can comprise from 0 elements to infinity (actually, a huge, gigantic, cosmic amount of finite elements). The *amount* of elements in the set is always an integer number although the elements themselves can be fractions, decimal numbers, and even transcendental numbers. For example, sale volumes over the past week can read something like {23,457.55, 17,444.02, 8,995.34, 7,234.75, 21,567.99, 9,999.99, 11,234.55}. There are exactly 7 total elements in the set. What is the probability to *guesstimate* a sale over 10,000? There are 3 favorable cases; $p = 3/7$.

There is a lot of confusion in the field of sets, both in the academia and among laypersons. The *combination* and *permutation* are the most commonly used terms. Worse, people interchange *combination* and *permutation* without any restriction! Apparently, in the British system of education, *combination* and *permutation* are synonyms! We must establish order and discipline in this classroom! (Don't worry, lads! I am mild mannered!)

There are four distinct types of sets from the most inclusive to the least inclusive: exponents, permutations, arrangements, combinations.

The *exponents* represent the most inclusive of the four types of sets. The *exponent set* includes *permutations*. The *permutation set* includes *arrangements*. The *arrangement set* is inclusive of *combinations*.

The number sets (*numerical* or *numeric*) are the most important mathematically. We can substitute the numbers by alphanumerical elements, such as words, names, any strings of characters. In the case of the alphanumerical sets, mathematics works with the indices (indexes) of the respective elements.

An example of ***exponents*** (N = 3, M = 3): 111, 112, 113, 121, 122, 123, 131, 132, etc. (a total of 27 sets). The best-known examples are the drawings of *pick or daily lotteries*, especially in the United States.

An example of ***permutations*** (for N = 3): 1 2 3, 1 3 2, 2 1 3, 2 3 1, 3 1 2, 3 2 1 (6 elements: 1* 2 * 3). There aren't many practical examples for permutations.

An example of ***arrangements*** (for N = 3, M = 2): 1 2, 1 3, 2 1, 2 3, 3 1, 3 2 (6 elements in set: 3 * 2).The best-known examples are the trifectas, the results of horse races (trifectas in the United States, triactors in Canada, top-three finishers in Britain, etc.)

An example of ***combinations*** (for N = 3, M = 2): 1 2, 1 3, 2 3 (3 elements: 3 * 2 / 1 * 2). The best-known examples are the drawings of *lotto games* conducted all over the world.

2. Exponents (Exponential Sets or Saliusian Sets; Ion Saliu Sets)

The ***exponential sets*** had been neglected by mathematics although they are the most important! All other sets are derived from exponents. Some number lovers (and teasers of mine!) call the exponents the *Ion Saliu sets*. I accept the nomination with grace and honor!

The exponents are very important. They are capable of solving a wide range of probability problems. Since the *exponents* accept both unique elements and duplicates (repeat elements), they can solve problems of gigantic proportions and importance. For instance, can intelligent life, as present on earth, have a duplicate anywhere in the gigantic universe? We will tackle the issue later in this book.

The pick-3 or pick-4 lottery games are the most commonly known examples of exponents. Each digit of the pick games takes values between 0 and 9 (10 values). The pick-3 game has a total of 10 to the power of 3 (10 ^ 3) or 1000 combinations. Examples: 013 (equivalent to 0, 1, 3 or 0-1-3).

The soccer pools, such as totocalcio (in Italy), have 3 outcomes for 13 games; 3 to the power of 13 (3 ^ 13) or 1,594,323 possibilities (total possible cases). Examples: 1,X,2,X,X,2,2,1,1,1,X,1,2 or 1X2XX22111X12.

The parameters for soccer pools (known as totocalcio, pronosport, or other names from nation to nation) are 13 (items per set), 0 (lower bound), 2 (upper bound). The no. 1 represents home victory, 2 is for win for the visitor, and 0 (marked as X on play slips) stands for a tie. The pick-3, pick-4 games have a lower bound of 0 and an upper bound of 9. Items per set are 3 (in pick 3) and 4 (in pick 4).

The formula of exponents is as follows:

Exponent (N, M) = N ^ M = N^M

(^ represents the *raising to power* operator; they also use the superscript as indicator of *raising to power* or *exponent, exponential operator*).

The lottery in the State of Pennsylvania has a digit game named Quinto. It draws 5 digits (**M** = 5**)** from 0 to 9 (a total of ten elements; **N** = 10). $\mathbf{10^5 = 100{,}000}$**.**

The exponents (exponential sets) grow much faster than anything else, including permutations!

3. Permutations (Factorial Sets, Exponents without Duplicates)

The ***permutations*** are also known as ***factorial*** as far as calculation is concerned.

The factorials also grow extremely rapidly, but with a lesser intensity compared to the exponents.

The formula of permutations is as follows:

Permutation (N) = N! = 1 x 2 x 3 x . . . x N

As I said previously, it's hard to find real-life examples of permutations. We can visualize an example by taking all the playing cards in a suit. There are 13 cards from 2 to A: 2, 3, 4, 5, 6, 7, 8, 9, 10, J, Q, K, A. We can arrange the 13 playing cards in a variety of orders without removing any card. Factorial of 13 or 13: 1 x 2 x 3 x 4 x 5 x 6 x 7 x 8 x 9 x 10 x 11 x 12 x 13 = 6,227,020,800 (more than 6 billion ways!). The first way is 2, 3, 4, 5, 6, 7, 8, 9, 10, J, Q, K, A; the next sequence is 2, 3, 4, 5, 6, 7, 8, 9, 10, J, Q, A, K; the third sequence is 2, 3, 4, 5, 6, 7, 8, 9, 10, J, A, Q, K; etc.

4. Arrangements Sets (*n* Permutations *m*)

The ***arrangements*** of *N* elements taken *M* at a time represent, in fact, a *partial permutation*. Older books use the notation ***nPm*** (***n Permutation m***) for arrangements. We can delve deeper mathematically.

nPm = n! / (n – m)!

n! = 1 x 2 x 3 x . . . x (n – m) x (n – m + 1) x (n – m + 2) x . . . x n

(n – m)! = 1 x 2 x 3 x . . . x (n – m)

We can simplify the fraction **n! / (n – m)!** and obtain

(n – m + 1) x (n – m + 2) x . . . x n

It is the factorial of a series of elements beginning at the term ***n – m + 1*** and ending at the term ***n***. That's the formula of the arrangements:

Arrangements (N, M) = N x (N – 1) x (N – 2) x (N – 3) x . . . x (N – M + 1)

The *exactas* (top two finishers) or *trifectas* (top three finishers) or *superfectas* (top four finishers) in horse racing are some of the most common representations of the arrangements.

If the famous horse race known as the Kentucky Derby has 20 horses, the total possible amount of straight trifectas is A (20, 3) = 20 x 19 x 18 = 6840.

The boxed trifectas represent, in fact, the ***combinations*** of (**N, M**). (20 x 19 x 18) / (1 x 2 x 3) = 6840 / 6 = 1140 *boxed trifectas*. See next.

5. Combinations (Boxed Arrangements)

The ***combination set*** is the best-known element of the four mathematical sets. The lotto drawings are some of the most common representations of the combinations. The random number generation of the arrangements is actually closer to the lotto drawings. The lotto numbers are not drawn in sequential order, but in sequences like 33, 7, 18, 44, 29, 48. The random number generation of the combinations in PermuteCombine.exe is like the lotto draws after sorting the numbers in ascending order.

The ***combinations*** are the equivalent of *boxed arrangements*. We sort the arrangements in sequential (lexicographic) order; therefore, the *order* of the elements is *not* important.

The combinations formula is as follows:

Combinations (N, M) = Arrangements (N, M) / Permutations (M) = {N x (N – 1) x (N – 2) x (N – 3) x . . . x (N – M + 1)} / {1 x 2 x 3 x . . . x M}

As in this example of the most common lotto game in the world: 6 from 49.

Total possible 6/49 combinations = (49 x 48 x 47 x 46 x 45) / (1 x 2 x 3 x 4 x 5 x 6) = 13,983,816.

The ***combination set*** is, by far, the most extensively researched. I reckon it is so because of the lotto games! Even members of the academia want to get rich. No kidding! A group of professors and staff members at Bradford University and College in Britain won the lotto jackpot! It happened on October 21, 2006, in the United Kingdom National Lottery. They played a lotto strategy I started in the early 1990s. I will touch that lotto strategy and also present the corresponding software later in this book.

People go a whole lot deeper in calculating a variety of probabilities regarding the combinations. We already saw several cases in the first chapter, especially the usage of the hypergeometric distribution probability formula. Lottery players apply so-called *lotto wheels* (abbreviated lotto systems). They don't play just one ticket of six numbers. They have a *pool* (group) of favorite lotto numbers, say 18.

What are the odds to get exactly one ticket with 4 of 6 winning numbers if the 6 winners come from the pool of favorite 18? I run my program *OddsCalc.exe*. The answer: the probability is 1 in 9.83.

Even deeper, the odds calculated as *at least* for 4 of 6 in 18 from 49 are 1 in 8.19 (just a slight difference).

6. Lexicographical Order: Each Element in the Set Has a Rank, Index, or Numeral Order

We know very well now how to calculate all possible elements in every type of numerical sets. We can also write software to generate all possible elements in every type of sets. I am the author of such incredible software: *PermuteCombine.exe*. It is well-known all over the world, especially for its uniqueness. It is the universal calculator and generator for exponents, permutations, arrangements, and combinations.

The generation can be set for any numbers or words. As of this time of writing (2009), no other piece of software can do what *PermuteCombine.exe* can perform.

So we want to generate all combinations in that lotto game where they draw 6 winning numbers from a field of 49. We can set a program such as *PermuteCombine.exe* to generate all possible combinations in the game (set). If the program is well-written and accurate, it should generate 13,983,816. The generating process will start with this typical combination: 1, 2, 3, 4, 5, 6. The generating will end with this combination: 44, 45, 46, 47, 48, 49.

We can see that the combinations are generated sequentially, or in lexicographic (lexicographical) order, from the first sequence to the last.

There are situations when generating all the elements in a set and counting them and then looking for a particular element is not an efficient process. We can see very easily what the first element in a combination set is without complex calculations or algorithms.

The combination 1-2-3-4-5-6 comes to mind automatically in the case of a lotto 6-from-49 game (any 6-number lotto game, actually). This is also easy: What is the combination of lexicographical order (or index) 13983816 in a lotto 6-of-49 game? Answer: 44, 45, 46, 47, 48, 49.

But how about index (or numeral orders) such as 77889 or 1000000 or 6991908? The index no. 6,991,908 is right in the middle of the set. It is represented by the lotto 6-49 combination 6-7-16-20-28-47.

The calculations are instantaneous with my own program *LexicographicSets.EXE*. The software is founded on some known algorithms released in the public domain. The first algorithm applied to the numerical sets known as combinations. The lexicographic-order algorithm was developed by *B. P. Buckles and M. Lybanon* to determine the rank (index) of a combination. The algorithm is included in the *Association for Computing Machinery* (ACM algorithm no. 515, published in 1977).

I discovered the opposite algorithm: determine the combination when the index (lexicographical order) is given.

Then I applied both types of algorithms to all four types of sets: exponents, permutations, arrangements, and combinations. I developed the combinations sets to further dimensions by creating lexicographic algorithms for *two-in-one* phenomena (such as *Powerball* lotto). To this date and my best knowledge, I am the only author of algorithms for lexicographic ordering for all four numeric sets.

Publishing and analyzing the algorithms are tasks beyond the scope of this book. Again, my Web site is open for business, including in this field. A simple search would lead to many resources at SALIU.COM, including the one-of-a-kind software (nowhere else to be found).

The lexicographical order is very important, however. We notice the case of lotto games where most combinations appear to be *truly random* to laypersons. They assess that a combinations like 6-7-16-20-28-47 appears to be *truly random*. On the other hand, the *infamous* combination 1-2-3-4-5-6 doesn't appear to be *truly random*; it appears to be strongly ordered.

In fact, the determining factor is the ***standard deviation***. I call ***standard deviation*** the *watchdog of randomness*. The extremes of the set (the beginning and the end) have combinations with very low standard deviations. Meanwhile, combinations with higher lexicographic orders (ranks, indexes) come from the inside of the set; their standard deviation is closer to the median. The common perception is that the higher the standard deviation, the *more random* a combination is!

Yes, there is a whole lot more on this topic at SALIU. COM. My software allows me to analyze huge amounts of data accurately and quickly. I analyzed real-life lottery drawings, and I determined beyond reasonable doubt that a low-standard-deviation combination such as 1-2-3-4-5-6 has a far lesser chance to hit high lotto prizes. In other words, it would require significantly more lottery drawings to achieve results similar to combinations with higher standard deviations.

This chapter afforded us to gain a lot of useful information in determining the total possible elements in numeric sets. That parameter is of the essence in calculating the probability.

Chapter III

THE FUNDAMENTAL FORMULA OF GAMBLING (FFG)

1. Theory of Probability Leading to the Fundamental Formula of Gambling (FFG)

Let's suppose I play the 3-digit lottery game (Pick 3, Daily 3, etc.) The game has a total of 1,000 straight sets. Thus, any particular pick-3 combination has a probability of 1 in 1,000 (we write it 1/1000). I also mention that all combinations have an equal probability of appearance. Also important—and contrary to common belief—is that the past draws do count in any game of chance, and Pascal demonstrated that hundreds of years ago. Evidently, the same-lotto-game combinations have an equal probability, *p*—always the same—but they appear with different statistical frequencies. Standard deviation plays an essential role in random events, as we just saw in the previous chapter. TheEverything, that is, for everything is random.

Most people don't comprehend the concept of all-encompassing *randomness* because phenomena vary in the particular probability, *p*, and specific degree of certainty, *DC*, directly influenced by the number of trials, *N*. Please read an important article at SALLIU.COM: "Combination '1, 2, 3, 4, 5, 6': Probability and Reality." A 6-number lotto combination such as 1-2-3-4-5-6 should have appeared by now at least once, considering all the drawings in all lotto-6 games ever played in the world. It hasn't come out and will not appear in my lifetime, I bet, even if I live 100 years after 2060, when Isaac Newton calculated that the world would end based on his mathematical interpretation of the Bible! (Newton and Einstein belong to the special class of the most intelligent mystics in human and natural history.) Instead, other lotto combinations, with a more *natural* standard deviation, will repeat in the same frame of time.

As soon as I choose a combination to play (for example 2-1-4), I can't avoid asking myself: *"Self, how many drawings do I have to play so that there is a 99.9% degree of certainty my combination of 1/1,000 probability will come out?"*

My question dealt with three elements:

- ***Degree of certainty*** that an event will appear, symbolized by ***DC***
- ***Probability*** of the event, symbolized by ***p***
- ***Number*** of trials (events), symbolized by ***N***

I was able to answer such a question and quantify it in a mathematical expression (logarithmic) I named the ***Fundamental Formula of Gambling (FFG)***. We already met the FFG in chapter 1. Indeed, the formula is fundamental.

$$N = \frac{\log(1 - DC)}{\log(1 - p)}$$

Indeed, *FFG* is the most essential formula of theory of probability. This formula was directly derived from the most fundamental formula of probability: number of favorable cases, *n*, over total possible cases, *N*: *n* / *N*.

The Fundamental Formula of Gambling (FFG) is a historic discovery in theory of probability, theory of games, and gambling mathematics. The formula offers an incredibly real and practical correlation with gambling phenomena. As a matter of fact, ***FFG*** is applicable to any sort of highly randomized events: lottery, roulette, blackjack, horse racing, sports betting, even stock trading.

By contrast, what they call *theory of games* is a form of vague mathematics: the formulas are barely vaguely correlated to real life.

2. Mathematics Leading to the Fundamental Formula of Gambling

I rationalized in this manner. The probability of any 3-digit combination is 1/1000. Therefore, I had expected that the repeat (skip) median of a long series of pick-3 drawings would be 500. It would be similar to coin tossing where the median of p = 1/2 series is 1. In other words, the median of a long series of coin tosses is 1.

To my surprise, the repeat median of long series of pick-3 drawings was not 500. It was closer to 700. I checked it for series of 1,000 real drawings and also randomly generated drawings. Then I checked the median against series of 10,000 (10 thousand) drawings. The median of the skip was always close to 700. Do not confuse it for the median combination in the set. That value of the median is, in fact, either 499 or 500. The correct expression is 4, 9, 9 or 4-9-9 or 5, 0, 0 (three separate digits).

What is that *median* useful for anyway? Among other properties, the *skip median* (or median skip) shows that, on the average, a pick-3 combination hits in a number of drawings. Any pick-3 combination hits within 692 drawings in at least 50% of the cases. Equivalently, if you play one pick-3 combination, there is a 50% plus chance it will hit within 692 drawings, or it will repeat no later than 692 drawings. The chance is also (almost) 50% that you will have to wait more than 692 drawings for your number to hit.

I studied theory of probability (gambling mathematics too!) in high school and in college. Some things get imprinted on our minds. Such information becomes part of our axioms. An axiom is a self-evident truth, a truth that does not necessitate demonstration. We operate with axioms in a manner of automatic thinking.

So I was analyzing mathematically long pick-3 series where $p = 1/1000$. Next, I wrote the probability of a single pick-3 number to hit two drawings in a row: $p = 1/1000 \times 1/1000 = 1/1000000$ (1 in 1 million). I have never found it useful to work with very, very small numbers in probability.

How about the reverse? The probability of a particular pick-3 number NOT to hit is $p = (1 - 1/1000) = 999/1000 = .999$. This is a very large number. It is almost certain that my pick-3 combination will not hit the very first time I play it.

How about not hitting two times in a row?

$P = (1 - 1/1000)^2 = 0.999$ to the power of 2 = .998. Still a very large number, mind you. I reversed the approach one more time: what is the opposite of not hitting a number of consecutive drawings? It is winning within a number of consecutive drawings.

The knowledge was inside my head. Unconsciously, I used Socrates' dialectical method of "delivering" the truth. (His mother delivered babies.) I also followed steps in De Moivre formula. At this point, I had this relation:

$$\mathbf{1 - (1 - p)^N}$$

where *N* represents the number of consecutive drawings.

I thought that for *N = 500* drawings, the expression above should give the *median* or a probability of *50%*. So I calculated 1 – (1 – p) N = 1 – (1 – 1/1000) 500 = 1 – (0.999) 500 = .3936 = 39.36%. Thus, my relation became

$$.3936 = 1 - (1 - 1/1000)^{500}$$

I made N = 692. I obtained the value

$$1 - (1 - 1/1000)^{692} = 1 - .5004 = 49.96\% \text{ (very close to 50\%).}$$

Next step, I made N = 693. I obtained the degree of certainty

$$1 - (1 - 1/1000)^{693} = 1 - .4999 = 50.01\% \text{ (very, very close to 50\%).}$$

Thus, the parameter I call the *FFG median* is between 692 and 693 for the pick-3 lottery.

I concluded I should not make more assumptions. What if I don't think I know what *N* should be for the median (50%), or for any other chance, which I simply called the degree of certainty? I realized I had the liberty to select whatever degree of certainty I wanted to and only had to calculate *N*. The relationship became

$DC = 1 - (1 - p)^N$

Then

$(1 - p)^N = 1 - DC$

The equation can be solved using *logarithms*:

$$N = \frac{\log(1 - DC)}{\log(1 - p)}$$

The only unknown is *N*, the number of consecutive drawings (or trials) that an event of probability, *p*, will appear at least once with the degree of certainty, *DC*.

The rest is history. I called the relation *the Fundamental Formula of Gambling* automatically. Unintentionally, it might sound cocky. Just refer to it as ***FFG***.

Nothing comes with *absolute certainty*, but to a *degree of certainty*! That's mathematics, and that's the only ***truth***.

In the year 2001, my memory dug out a real gem. I wrote about it in an article on my Web site: "Cool Stories of the Truth."

I found another treasure, a little book in Romanian. Don't they say great things come in small packages? It couldn't be truer than in this case. The book was (***Les certitudes du hasard*** *[The Certainties of Hazard]*) by French academician Marcel Boll. The book was first published in French in 1941. My 100-page copy was the 1978 Romanian edition. It all came to life like awakening from a dream. The book presented a table very similar to the table on my Fundamental Gambling

Formula page. Then, in small print, the footnote: *"The reader who is familiar with logarithms will remark immediately that N is the result of the mathematics formula: N = log(1 – pi) / log(1 – p)."*

That's what I call the Fundamental Formula of Gambling indeed! Actually, the author *Marcel Boll* did not want to take credit for it. *Abraham de Moivre* largely developed the formula. Then I remembered more clearly about de Moivre and his formula from my school years. *Abraham de Moivre* himself probably did not want to take credit for the formula. As a matter of fact, the relation only deals with one element: the probability of *N* consecutive successes (or failures). Everybody knows, that's $\mathbf{p}^{N}$ (*p* raised to the power of *N*). It's like an axiom, a self-evident truth. Accordingly, nobody can take credit for an axiom. I thought *Pascal* deserves the most credit for establishing *p* = *n* / *N*. From there, it's easy to establish *p* ^ *N*. And give birth to so many more worthy numerical relations.

3. The Fundamental Table of Gambling (FTG)

Substituting *DC* and *p* with various values, the formula leads to the following very meaningful and useful table. You may want to keep it handy and consult it, especially when you want to bet big (as in a casino).

Number of Trials, N, *Necessary for an Event of Probability,* p, *to Appear with the Degree of Certainty,* DC

DC	P= .90	p= .75	p= .66	*p= 1/2*	p= 1/3	p= 1/4	p= 1/6	p= 1/10	p= 1/16	p= 1/32	p= 1/64	p= 1/100	p= 1/1000
10%	—	—	—	—	—	—	—	1	1	3	6	10	105
25%	—	—	—	—	—	1	1	3	4	9	18	28	287
50%	*1*	*1*	*1*	*1*	*1*	*2*	*3*	*7*	*10*	*21*	*44*	*68*	*692*
75%	1	1	2	*2*	3	4	7	13	21	43	88	137	1,385
90%	1	2	2	*3*	5	8	12	22	35	72	146	229	2,301
95%	1	2	3	*4*	7	10	16	29	46	94	190	298	2,994
99%	2	3	4	*7*	11	16	25	44	71	145	292	458	4,602
99.9%	3	5	6	*10*	17	24	37	66	107	217	438	687	6,904

Let's try to make sense of these numbers. The easiest to understand are the numbers in the column under the heading "***p = 1/2***." It analyzes the coin-tossing game of chance. There are 2 events in the game: *heads* and *tails*. Thus, the individual probability for either event is *p = 1/2*. Look at the row "50%": it has the number 1 in it. It means that it takes 1 event (coin toss, that is) in order to have a fifty-fifty chance (or degree of certainty of 50%) that either heads or tails will come out. More explicitly, suppose I bet on heads. My chance is 50% that heads will appear in the first coin toss. The chance or degree of certainty increases to 99.9% that heads will come out within 10 tosses!

Even this easiest of the games of chance can lead to sizable losses. Suppose I bet $2 before the first toss. There is a 50% chance that I will lose. Next, I bet $4 in order to recuperate my previous loss and gain $2. Next, I bet $8 to recuperate my previous loss and gain $2. I might have to go all the way to the ninth toss to have a 99.9% chance that, finally, heads will come out! Since I bet $2 and doubling up to the ninth toss, 2 to the power of 9 is 512. Therefore, I needed $512 to make sure that I am very, very close to certainty (99.9%), that heads will show up and I win $2!
Very encouraging, isn't it?

Actually, it could be even worse: it might take 10 or 11 tosses until heads appear! This dangerous form of betting is called a Martingale system. You must know how to do it—study this book thoroughly and grasp the new essential concepts: ***Number of trials,*** **N,** and especially the ***Degree of Certainty,*** **DC** (in addition to the ***probability,*** **p**). Most people still confuse probability for degree of certainty or vice versa. Probability in itself is an abstract, lifeless concept. Probability comes to life as soon as we conduct at least one trial. The probability and degree of certainty are equal for one and only one trial (just the first one ever!). After that quasi-impossible event (for coin tossing has never been stopped after one flip by any authority), ***the degree of certainty,* DC, *rises with the increase in the number of trials,* N, *while the probability,* p, *always stays constant*.** No one can add faces to the coin or subtract faces from the die for sure and undeniably. But each and every one of us can increase the chance of getting heads (or tails) by tossing the coin again and again (repeat of the trial).

Normally, though, you will see that heads (or tails) will appear at least once every 3 or 4 tosses (the *DC* is 90% to 95%). Nevertheless, this game is too easy for any player with a few thousand dollars to spare. Accordingly, no casino in the world would implement such a game. Any casino would be a guaranteed loser in a matter of months! They need what is known as *house edge* or *percentage advantage*. This factor translates to longer losing streaks for the player, in addition to more wins for the house! Also, the casinos set limits on maximum bets: the players are not allowed to double up indefinitely.

A few more words on the house edge. The worst type of gambling for the player is conducted by state lotteries. In the digit lotteries, the state commissions enjoy typically an extraordinary 50% house edge! That's almost 10 times worse than the American roulette—considered by many as a suckers' game! (But they don't know there is more to the picture than meets the eye!)

In order to be as fair as the roulette, the state lotteries would have to pay $950 for a $1 bet in the 3-digit game. In reality, they now pay only $500 for a $1 winning bet! Remember, the odds are 1,000 to 1 in the 3-digit game. If private organizations, such as the casinos, would conduct such forms of gambling, they would surely be outlawed on the grounds of extortion! In any event, the state lotteries defy all antitrust laws: they do not allow the slightest form of competition! Nevertheless, the state lotteries may conduct their business because their hefty profits serve worthy social purposes (helping the seniors, the schools, etc.). Therefore, lotteries are a form of taxation—the governments must tell the truth to their constituents.

4. The Fundamental Formula of Gambling: Games other than Coin Tossing

Dice rolling is a more difficult game, and it is illustrated in the column "*p=1/6*." I bet, for example, on the 3-point face. There is a 50% chance (degree of certainty, *DC*) that the 3-point face will show up within the first 3 rolls. It will take, however, 37 rolls to have a 99.9% certainty that the 3-point face will show up at least once. If I bet the same way as in the previous case, my betting capital should be equal to 2 to the power of 37! It's already astronomical, and we are still in easy-gambling territory!

Let's go all the way to the last column, "*p = 1/1000*." The column illustrates the well-known 3-digit lottery game. It is extremely popular and supposedly easy to win. Unfortunately, most players know little, if anything, about its mathematics. Let's say I pick the number 2-1-4 and play it every drawing. I only have a 10% chance (*DC*) that my pick will come out winner within the next 105 drawings! The degree of certainty, *DC*, is 50% that my number will hit within 692 drawings! Which also means that my pick will not come out before I play it for 692 drawings. So I would spend $692, and maybe I would win $500! If the state lotteries want to treat their customers more fairly, they should pay $690 or $700 for a $1 winning ticket. That's where the *fifty-fifty-chance* line falls. In numerous cases, it's even worse. I could play my daily-3 number for 4,602 drawings and, finally, win. Yes, it is almost certain that my number will come out within 4,602 or within 6,904 drawings! Real-life case: Pennsylvania State Lottery has conducted over 6,400 drawings in the pick-3 game. The numbers 2, 1, 4 have not come out in the first 6,400 drawings.

All lottery cases and data do confirm the theory of probability and the formula of bankruptcy—I mean, of gambling! By the way, it is almost certain (99.5% to 99.9%) that the numbers 2-1-4 will come out within the next 400–500 drawings in Pennsylvania lottery. But nothing is 100% certain, not even 99.9999%!

We don't need to analyze the *lotto* games. The results are, indeed, catastrophic. If you are curious, simply multiply the numbers in the last column by 10,000 to get a general idea. To have a 99.9% degree of certainty that your lotto ticket (with 6-from-49 numbers) will come out a winner, you would have to play it for over *69 million* consecutive drawings! At a pace of 100 drawings a year, it would take over *690,000 years*!

5. The Practical Dimension of the Fundamental Formula of Gambling

The Fundamental Formula of Gambling does not explicitly or implicitly serve as a *gambling system*. It represents pure mathematics. Users who apply the numerical relations herein to their own gambling systems do so at their risk entirely. I, the author, do apply the formula to my *gambling* and *lottery* systems. I will show you how to use the gambling formula, my software application *MDIEditor and Lotto WE* and the *lotto* systems that come with the big program. I will put everything in a winning *lotto strategy* that targets the third prize in lotto games (4 out of 6).

At later times, I also released gambling systems, strategies for *roulette, blackjack, baccarat, horse racing, sports betting*. Is it all? Probably NOT! You'll find some more at SALIU.COM. But firstly, I will present and analyze several strategies and systems for lottery and gambling in this very book. Again, it is pure mathematics first and foremost. The theories, however, prove to have undeniable application in real-life phenomena. Add to all this corresponding *software* to make it much easier to devise systems and generate winning outcomes. The software can be easily obtained by becoming a registered member of SALIU.COM for a reasonable fee (valid for the existence of my Web site).

Can the Fundamental Formula of Gambling have business applications? You bet! Think of *stock trading* or *currency trading*. They are, like everything else, random phenomena. Granted human psychology plays a role in the outcome of stock trading or foreign-currency trading.

There are situations of *frenzy*, for example. A former chief of the U.S. Federal Reserve Board (FED), Allen Greenspan, called such situations *irrational exuberance*. It seems as if randomness is put aside in frenetic situations: Everybody buys and buys and buys, and the prices go higher and higher and higher.

There are also situations of *panic*. It happened in 2008 when the whole world was hit by an economic recession. Randomness is put aside in *panic* situations as well. Everybody sells and sells and sells, and the prices go lower and lower and lower.

Randomness still rules nevertheless. The markets and the economies come back to balanced states of *normal randomness*.

There is significant randomness in stock evolution or currency exchange. Many stockbrokers came to terms with the reality that all stocks fluctuate in an undeniably random fashion. I am surprised how many brokerage firms have visited my Web site!

6. The Profound Implications of the Fundamental Formula of Gambling

Aren't the computers amazing? I wrote software to handle the Fundamental Formula of Gambling (FFG) and its reverse: Anti-FFG or the Degree of Certainty. There are situations when we want to calculate the Degree of Certainty that an event of probability, *p*, will appear at least once within a number of trials, *N*. As a matter of fact, this method offers a more precise correlation between an integer number of trials and a degree of certainty, *DC*, expressed as a floating-point number. Furthermore, the program can determine the probability from a data series! The number of elements in the data series is known (*N*). Sorting the data series can determine the median: the degree of certainty, *DC*, equal to 50%!

SuperFormula.exe calculates several mathematical, probability, and statistics functions: binomial distribution; standard deviation; hypergeometric distribution; odds (probability) for lotto, lottery, and gambling; normal probability rule; sums and mean average; randomization (shuffle); etc.

The program name is *SuperFormula.exe*, 32-bit character-mode software for DOS/Windows. *SuperFormula.exe* also calculates the Binomial Distribution Formula, the Binomial Standard Deviation, and then some.

There appears to be a *glitch* in the software. We see in the *Fundamental Table of Gambling (FTG)* that the degree of certainty, *DC*, goes all the way up to 99.9%. We can use

SuperFormula.exe for higher degrees of certainty: 99.99%, 99.999%, 99.9999%, etc. How about 100%? We want something, like a gambling system, that offers a 100% degree of certainty. Oops! That's where the *glitch* is triggered, many software users cried!

Actually, it is not a glitch. The occurrence gave me the opportunity to discover the most essential attribute of the Universe: randomness. It is all random. Nothing is *absolutely certain.* Everything, the Universe including, comes in a degree of certainty or another. We can write a very long string of the degree of certainty, something like 99.99999999999999999 99999999999999999999999999% . . . The length of a page or the length of an infinite number of pages.

But it can **never** be 100%—ever!

I put it humorously:

This is why God fears mathematics, while Einstein hates gambling!

Abraham de Moivre, a French/English refugee mathematician and philosopher, discovered the first steps of this formula that explains the *Universe* the best. I believe M. de Moivre was frightened by the implications that finalizing such formula would have led to ***the absurdity of the concept of God***. I did finalize the formula for the risks in my lifetime pale by comparison to the eighteenth century.

Absolute certainty represents the main attribute of God.

Thusly, God, no doubt, represents the limit of mathematical absurdity and, therefore, of all absurdity. And thusly, we discovered here the much-feared mathematical concept of *Degree of Certainty, DC*. I introduced the DC concept in the year of grace 1997 or 1997 + 1 year after *tribunicia potestas* were granted to Octavianus Augustus (the point in time humans started the year count of *Common Era* or *CE*, which is still in use today).

The Internet search on degree of certainty yielded one and only one result in 1998: my Web site (zero results in 1997 for DC was introduced in December of that glorious year, with some beautiful snowy days just before the Global Warming debate started). For we shall always be mindful that nothing comes in absolute certainty; everything comes in degrees of certainty—never zero, never absolute. It is all a matter of probability, not absolute certainty.

My life's work will present these most profound implications of FFG in a dedicated book.

Chapter IV

STANDARD DEVIATION: A MEASURE OF RANDOMNESS

1. Introduction to the Concept of Standard Deviation

You already learned in previous chapters that *randomness* is the essence of the *Universe*. ***Standard deviation*** is a most fundamental element of randomness. This chapter offers a most thorough analysis of standard deviation. You might have not gotten the chance to read the best materials on standard deviation. Apparently, the search engines favor Web pages with very short content. You have the chance now to get to the most relevant and thorough writing on standard deviation, variance, volatility, fluctuation, data dispersion, and the likes.

Plus, you'll get to acquire the best software in the field that, by the way, is freeware! You can calculate the standard deviation of millions of terms or elements in a data series. You no longer need to hassle with *imprecise* concepts such as *pooled standard deviation* or *sample standard deviation*. You'll be able also to best evaluate the standard deviation. Is the data series uniform, or is the data too dispersed (volatile)? The size of the *standard deviation* in a data series must be correlated to other statistical parameters, notably the *mean average* and especially the *median*. A data series with a standard deviation less than the median is more desirable as it is more highly predictable in its future movements.

It started in the American culture with the Computers for Dummies book series. I strongly resisted the concept in the very beginning. Then I realized we all are dummies in most fields of knowledge. If *Socrates* was right in one and only one thing, it is his assertion "I know that I know nothing." I was a dummy in the field of statistics and standard deviation. Life made me acquire knowledge to handle frequent problems that can only be solved by knowing, beyond *dummyness*, the deviation in its standard mathematical expression.

So you, the educated ones in standard deviation, don't curse me for moments that look like a "standard deviation for dummies" treatment of the topic. Matter of fact, you'll feel like you were a dummy as well, as far as the evaluation of the standard deviation was concerned. You didn't know that before reading this material. It took me years of gambling experiences to come to a *clear mathematical evaluation of the standard deviation. The standard deviation of an event (materialized in a data series) is desirable to be under three (3) times the median of the series.*

The *median* is a very essential element in randomness and statistical data. I discovered that the *median* represents the *number of trials*, N, for a *degree of certainty*, DC, equal to *50%*. It is the threshold where possible and impossible are equal. The standard deviation is intimately related to the median, number of trials, and degree of certainty.

Many probability formulas or assessments can be validated with higher degrees of certainty when the standard deviation is below three times the median of the result series. Everything that probability theory says about coin tossing, or throwing the dice, is validated by the standard deviation of mathematically big number experiments.

Also, if data doesn't fit within three standard deviations, forget about it! The formula or assessment isn't mathematically valid. The law of big numbers isn't actually that big. The big numbers are benignly low. Nothing like that scary infinity! Read more deeply *Mathematics of Fundamental Formula of Gambling* and *Ion Saliu's Paradox of N Trials* at SALIU.COM.

In some cultures, the name for variance is average squared deviation or average of squared deviations. The standard deviation becomes average deviation or average of deviations, the absolute (unsigned) values of the deviations being considered. The best measure of random variation, however, is the FFG deviation. You may have the chance to learn more about it in a future book or a Web page of mine. I do not yet reveal it because it is part of my gambling strategy that I use in casinos.

The problem with the standard deviation is its huge fluctuation. One and only one term of the series can have an unreasonably big impact on standard deviation. That's why, in some judgmental sports, they throw out the lowest score and the highest score before calculating the average. It's one simple way of controlling the bias. Moreover, the standard deviation does not indicate the future direction (the short-term trend). Will the next term be higher than the most recent element of data—or will we record a decrease?

This author offers this type of evaluation for statistical data as related to the standard deviation.

• The data series is uniform (less dispersed, spread) therefore easier to analyze and control, if its standard deviation is less than or equal to the mean average and especially the **median**.

• The data series is volatile (highly dispersed, spread) therefore more difficult to analyze and control, if its standard deviation is three times greater than the median and especially the **mean average**.

2. The Binomial Standard Deviation Formula (BSD)

The ***binomial standard deviation*** applies to events with two outcomes: *win* or *loss*. For example, betting on *heads* in coin tossing can lead to win (the appearance of heads) or loss (the appearance of the opposite, *tails* in this case).

The binomial standard deviation is calculated by the following formula:

Standard Deviation = Square_Root{(N * p * (1 – p)}

That is, standard deviation equal to the square root of the number of trials (events), ***N***, multiplied by the probability, ***p***, multiplied by the opposite (*complementary*) probability (or **1 minus *p***), where *SQR()* represents the square root function, *p* is the probability of appearance, and *N* symbolizes the number of trials.

Suppose we toss a coin 100 times (N = 100). The probability of heads is p = 1/2 = .5. The standard deviation is SQR{100 * .5 * .5} = SQR(100 * .25) = SQR(25) = 5. The expected number of heads in 100 tosses is .5 * 100 = 50.

This type of standard deviation leads to an essential rule in mathematics: ***normal probability rule***.

The rule of normal probability proves that in *68.2%* of the cases, the number of heads will fall within *one (1) standard deviation* from the number of expected successes (50). That is, if we repeat 1,000 times the event of tossing a coin 100 times, in 682 cases, we'll encounter a number of heads between 45 and 55.

The rule of normal probability proves that in *95.4%* of the cases, the number of heads will fall within *two (2) standard deviations* from the number of expected successes.

The rule of normal probability proves that in *99.7%* of the cases, the number of heads will fall within *three (3) standard deviations* from the number of expected successes.

That's what the *normal curve* or the *Gauss bell* is all about.

3. The Statistical Standard Deviation

There is no formula to calculate the statistics standard deviation directly (?). That's what they told you in school. That's what they say in other public places with the self-proclaimed goal of education. Only an algorithm can lead to the standard deviation of a data series, they say.

Indeed, the algorithm is always available. The following are the steps of the algorithm implemented in my free software *SuperFormula.exe*.

- Sum up data.
- Calculate the mean average (sum total divided by the number of elements).
- Deduct each element of the collection from the average. Raise each difference to the power of 2.
- Add up the squared differences.
- Divide the new sum total by the number of elements in the data series.

- The result represents the variance; the square root of the variance represents the famous standard deviation.

A data series like 1, 2, 3, 6 has a mean average (μ, *mu*) equal to

$\mu = (1 + 2 + 3 + 6) / 4 = 3$.

The differences from the mean are –2, –1, 0, +3. The variance (σ^2 [sigma squared]) is the measurement of the squared deviations. The variance is calculated as

$\sigma^2 = \{(-2)^2 + (-1)^2 + 0 + 3^2\} / 4 = 14 / 4 = 3.5$.

Finally, the standard deviation (σ, sigma) is equal to the positive square root of the variance:

$\sigma = SQR(3.5) = 1.87$.

Nevertheless, there are formulas (plural, indeed) to calculate the statistical deviation in advance. There is a dominant deviation parameter in all the stochastic (probabilistic) events. In fact, all events are stochastic since randomness is present in everything-there-is. Nothing-there-is can exist with absolute certainty (see the mathematics of the absurdity of absolute certainty in the previous chapter on FFG). The elements of a stochastic phenomenon deviate from one another following mathematical rules. The difference is in the probability of the event (phenomenon). The probability then determines subsequent parameters such as median, volatility, standard deviation, FFG deviation, etc.

In 2003, I announced on my Web site that I had discovered a formula for a very important measure in the fluctuation of probability events: *FFG deviation*. Soon thereafter, I have been bombarded with requests to present the formula for FFG deviation and the statistical standard deviation. Of course, I was asked (in strong terms sometimes) to release also free software to accompany the formula calculations. The requests were also presented in public forums, sometimes strongly worded.

At that time, I did not publish the formulas to calculate the FFG deviation and the statistical standard deviation. Such an act would have served people I do not want to serve. They belong to the following categories: gambling developers and high rollers, lottery systems and software developers, stock traders. I know exactly whom I am talking about. I have received many a message from them. They inundated my inbox with correspondence, and my postal box. They would be the ones that would charge serious money out of my effort. The vast majority of people do not really need to know exactly all the formulas involved in standard deviation calculations.

Suffice to say that my software does incorporate standard deviation calculations. Also, the greatest random number/ combination generator—*IonSaliuGenerator ActiveX control*—makes extraordinarily good usage of the standard deviation and deltas (absolute differences between two terms of a series).

This is a book in print. The copyright protection is much stronger than the e-book case (electronic format). So I can tell you now that *the **FFG deviation** represents the **number of trials,*** **N,** *for a **degree of certainty equal to 25%**.*

We can also calculate in advance the statistical standard deviation as half of the median of the mean average.

All my data analyses prove the two formulas above beyond any reasonable doubt. It is exactly like that. I saw it in roulette spins and in lotto drawings for thousands and thousands of real-life results.

4. The Best Software to Calculate or Determine Standard Deviation

At SALIU.COM, you can find and download potent software to do a multitude of calculations on the topic of *standard deviation*, plus *theory of probability* and *statistics*. Two programs stand out: *FORMULA.EXE* and *SuperFormula.exe*. FORMULA.EXE is 16-bit software, now superseded by ***SuperFormula.EXE***. The latter takes a data file (consisting of real numbers or simulations) and calculates the standard deviation, sum, mean average, median, minimum, and maximum. *SuperFormula.EXE* has dozens of functions for mathematics, statistics, probability, and combinatorics. Read the manual at SALIU.COM/formula.html: "*Thorough Analysis of Standard Deviation: Standard Deviation, Variance, Variability, Fluctuation, Volatility, Variation, Dispersion, Median, Mean Average.*"

The free software does not calculate or present directly the formulas for *FFG deviation* and the *statistics standard deviation*. But I have written a bundle of computer programs that do such calculations. The standard deviation formulas have been validated for millions of cases, both real-life and simulations. The cases cover very popular probability events such as lotto and lottery, roulette, horse racing, and yes, stocks! (If a company is not sick, the stock fluctuates very closely in accordance to the FFG deviation like roulette spins!)

Such software serves a small group of users with strong interests. Therefore, a fee must be in place regarding the acquiring of special-interest software. My free software download page presents the terms and conditions regarding custom software by Ion Saliu.

They are after software like *SUMS.EXE*. You can download the program from my software download site (on the footer of every Web page).

The application is extraordinarily powerful. It calculates meaningful statistics for a lotto/lottery data file: Sum, Root Sum (Fadic Addition), Average, Standard Deviation, Average Deviation from Mean Average, Average of Deltas. At the end of the report, *SUMS.EXE* calculates the medians of the above statistical parameters. And then they want a lot more: the formulas to calculate in advance the standard deviation, the average of the deviations from the mean, and the average of the deltas.

Here is a sample report for a statistically large data file in the Pennsylvania lotto 5/39 game.

Statistics for File: C:\LOTTERY\LOTTO 5-39
Total Lines: 3768

```
Drawings Sum Root Average StdDev AvgDev AvgDel

1-16-19-22-26 84 3 16.80 8.57 6.64 6.25
9-11-13-22-38 93 3 18.60 10.67 9.12 7.25
8-14-23-28-29 102 3 20.40 8.16 7.52 5.25
19-22-27-31-38 137 2 27.40 6.71 5.68 4.75
. . . .
Medians: 100 5 20.00 10.03 8.64 6.75
```

The three fundamental forms of dispersion (deviation):

StdDev = Standard Deviation
AvgDev = Average Deviation (disregarding the sign)
AvgDel = Average Deltas (absolute differences)

The average of the absolute deviations is always less than the standard deviation. In turn, the average delta is always less than the average deviation and much less than the standard deviation.

You can also see that the median of the *standard deviation* was very close to *half* the median of the *mean average*.

Analysis of real-life data must always back the formulas—or invalidate formulas. A relation is not a formula if invalidated by data analysis. A rule is not a rule mathematically if data proves just one exception. If things deviate from an established norm, they must do so in accordance to the rules of the watchdog of randomness: Standard Deviation.

5. Real-Life Case History: Standard Deviation and Politics

The U.S. presidential election of 2000 was some political event! The event had also probabilistic merit. Theory of probability can play a positive role in assuring political fairness.

There are situations when an election becomes a coin toss. That's normal, there is no problem in that. The problem is not to recognize a situation as such and handle it accordingly. We can see below a report for 100 million coin tosses. The standard deviation is only 5,000. Three times the standard deviation is 15,000. The president of the United States was declared winner by a margin of fewer than 500 votes in the state of Florida (where the governor was the brother of the new President!).

If a political candidate has no more than half the votes plus 3 standard deviations, the election fails the coin-toss test. It is possible that a new election will show the other candidate winning. But if a candidate wins the first election by more than 3 standard deviations, it is almost certain to declare that the election was not a coin toss.

In my opinion, a new election should always be set after any coin-toss situation. Now if the new election brings the same under-3 standard deviations result, the winner should deserve the benefit of the doubt. It is certain that most voters would deal with a second election more responsibly. Moreover, more of those who did not vote the first time would get involved the second time around.

The standard deviation for an event of probability p = 1/2 in 100,000,000 binomial experiments is
BSD = 5,000

The expected (theoretical) number of successes is 50,000,000.

Based on the Normal Probability Rule,

* 68.2% of the successes will fall within 1 Standard Deviation from 50,000,000 (i.e., between 49,995,000–50,005,000),
** 95.4% of the successes will fall within 2 Standard Deviations from 50,000,000 (i.e., between 49,990,000–50,010,000), and
*** 99.7% of the successes will fall within 3 Standard Deviations from 50,000,000 (i.e., between 49,985,000–50,015,000).

Again, the results above were calculated by *SuperFormula.exe* for 100 million cases. The electoral commission should have called for new elections, at least in the state of Florida.

Chapter V

THE BIRTHDAY PARADOX OR THE PROBABILITY OF REPETITION (COINCIDENCES, DUPLICATION, COLLISIONS)

---

1. Introduction to the Birthday Paradox: Probability of Coincidences, Repetition

The concept of *randomness* is not well understood at this time. There is a lot of ambiguity out there. Randomness means many things to many people. The precise presentation of randomness is one of the main goals of this book.

Everything in the Universe represents the cause-effect interaction of elements. The only form of interaction is represented by *randomness*. The opposite of randomness is *order*. Order always follows a *plan*. If there is a plan, there must be a *planner*. Only a *god* can be a planner at the scale of the entire Universe. But a god must pass the test of *absolute certainty*, as we learned in the presentation of FFG. Mathematics proves, however, that absolute certainty is *absolute absurdity*. Therefore, randomness is the only attribute of the Universe that is *valid* mathematically.

We saw in chapter 2 that any finite number of elements can be put together in groups based on certain rules. Such groups are known as sets. The sets can comprise from 0 elements to infinity (actually, a huge, gigantic, cosmic amount of finite elements).

The exponential sets (or Ion Saliu sets) represent the most comprehensive type of sets. Such sets can have both *unique* elements and *duplicates* (collisions or repeated elements). Everything that there is, was, or will be in the Universe can be placed in a cosmically gigantic exponential set. Each and every element comes into being randomly. The generation of the elements is the result of randomness, *not* the result of ordering or planning (for there is no planner!).

Only humans attempt to control, *relatively*, randomness. The humans (representatives of awareness or intelligent life) are the only element in the Universe to achieve *order*. Again, that order is valid only in relative terms. As an example, humans created computer programs that generate exponential sets in lexicographical order. The software created by humans can also generate numerical sets in random fashion, very much like everything that occurs in nature and the Universe.

If the generating process is random, the elements in the set will repeat with a degree of certainty precisely calculated. The probability of repetition is commonly known as the *Birthday Paradox*.

Warren Weaver makes a nice popularization of the *Birthday Paradox* in his famous book *Lady Luck* (p.132):

> Suppose there are N people in a room. What is the probability that at least two of them share the same

> birthday—the same day of the same month? . . . When there are 10 persons in a room together, this formula shows that the probability is 0.117 (11.7%) that at least two of them have the same birthday. For N = 22 the formula gives p = 0.476 (47.6%); whereas for n = 23 it gives p = 0.507 (50.7%) . . . Most people find this surprising. But even more surprising is the fact that with 50 persons, the probability is 0.970. And with 100 persons, the odds are better than three million to one that at least two have the same birthday.

The *Birthday Paradox* was born in 1938, however. It referred to fishes, to all creatures. But this is just one element of a much bigger picture, like the fish compared to the ocean. The complete picture has the caption: *The probability of duplication or collisions*.

2. Mathematics, Formula of the Birthday Paradox: The Saliusian Sets

The ***Birthday Paradox*** is just one particular case of ***exponential*** sets. They are also known as *Saliusian sets* or *Ion Saliu sets*. The exponents are sets of numbers where the duplicates are allowed. By contrast, the other mathematical sets consist of unique elements only (e.g., permutations or combinations). The exponential sets had been neglected by mathematics although they are the most important! All other sets are derived from exponents.

The *Birthday Paradox* resembles the pick lottery (or daily lotteries; e.g., pick 3), 1x2 soccer pools, and everything related to the sets known as exponents. My free software *PermuteCombine.exe* generates any type of sets, including exponents (1-1-1-2-2-3, 1-1-1-2-3-2, 1-2-3-4-5-6) and combinations (1-2-3-4-5-6, 1-2-3-4-5-7).

The *pick-3* game has the following parameters:

1) The LOWER bound = 0
2) The UPPER bound = 9
3) The number of ELEMENTS = 3

In this case of exponential sets, N = 3 and M = 9 – 0 + 1 = 10. Total possible elements: 3^{10} = 1000. The 1,000 elements of the *pick-3* set contain unique sets (1, 2, 3), plus double-digit combinations (1, 1, 0), plus triple-digit sets (9, 9, 9). The *unique* elements are easily calculated by using the *arrangements of (M, N)*: 10 * 9 * 8 = 720. Yes, it is the number of trifectas in horse racing for 10 horses.

If we deduct the *unique* elements from *total elements*, the result represents number of elements with *at least two digits being equal*. In the pick-3 case, there are 1,000 – 720 = 280 elements with *double* and *triple* digits. Obviously, the probability that *at least two digits are repeats* is 280 / 1000 = .28 or 28%.

The general formula of the Birthday Paradox (collisions, repetition, or duplication) is a two-step algorithm:

Number_of_Duplicate_Sets(M, N) = Exponents(M, N) – Arrangements (M, N)

Probability_of_Collisions (Coincidences or Birthday Paradox) = Number_of_Duplicate_Sets(M, N) / Exponents(M, N)

In the example for *365 birthdays*:

Exponents of (365, N) – Arrangements of (365, N)

where *N* represents the *number of persons* randomly chosen.

The interesting case is ***N*** **>** ***M***. Take as an example, *dice rolling*. Number of elements *M = 6* (6 numbered faces from 1 to 6). If we throw *2* dice at a time, the probability of *duplication* (e.g., 1-1 or 6-6) is 16.66% (*6 / 36*). If we throw *6* dice at a time, the probability of duplication (e.g., 1-1-?-?-?-? or ?-6-?-?-?-6) is 98.46% (*45,936 / 46,656*). If we throw *7* dice at a time (*7 > 6*), the probability of duplication (e.g., ?-?-?-?-1-?-1 or 6-?-?-?-?-6-?) is 100% (*279,936 / 279,936*).

In such cases, the algorithm sets the arrangements (number of elements with NO duplication) to zero by default. The algorithm proceeds to dividing Total number of sets (M ^ M) to the equal value of Number of sets with duplicates [(M ^ M) – 0]. It is not a case of mathematical absurdity. It is possible to throw seven dice at a time. But all trials will show *at least two faces being equal or duplicate*.

My software does exactly that by employing the correct (mathematically valid) algorithms.

3. The Best Software to Calculate the Probability of Collisions, Coincidences, Duplication

No doubt about it. It is very tedious, indeed, to perform all the calculations manually. I mean, even using a calculator, let alone via paper and pencil! And that's why we, the humans, write computer programs.

Yours truly has written unarguably the best software to tackle the hard calculations of the birthday paradox type (or collisions or duplication). Once again, such software can be had easily at my Web site SALIU.COM. My software is still unique of the time of this writing. I will refer you to two programs especially: *BirthdayParadox.exe* and *Collisions.exe*. This is the most accurate and comprehensive software for calculations in the probability of duplication (or collisions or coincidences), including the tiny Birthday Paradox. The software performs the calculations both ways:

1. Calculate the probability (degree of certainty) when the number of persons (elements) is known
2. Calculate the number of persons (elements) when the probability is known

Case 1 is better known, including the formula. Case 2 is far less known, especially the algorithm. My software is the *only* one that can automatically perform the calculations for case 2. I must say once more that presenting software algorithms is beyond the purpose of this book. Presentation of algorithms can be the object only of computer science or computer programming books.

The *birthday paradox* can only tell the number of duplicate sets, but not the structure of the sets. That is, 2-repeat elements are counted together with 3-repeat elements, plus 4 repeats, etc. My approach lets you determine beforehand the repeat sets by category. Look at the pick-lottery games. In the pick-3 game, M = 10, N = 3. The number of triple digit sets is equal to *M*(10, in this case). The number of exactly double-digit sets is equal to M * (M – 1) * N (270, in this case). In the pick-4 game, M = 10, N = 4. The number of quadruple digit sets is equal to *M*(10, in this case). The number of exactly triple-digit sets is equal to M * (M – 1) * N (360, in this case).

The user can generate the sets for the chosen parameters. The sets can be either in lexicographic order (all of them) or generate any amount of random sets. If total possible number of sets is greater than 10,000,000, the program warns *not* to generate lexicographical sets. It would gobble up a hard disk!

Collisions.exe works with far larger numbers than *BirthdayParadox.exe*. I employed all the programming tricks that I know to make possible calculations for huge numbers.

BirthdayParadox.exe works best with birthday cases (i.e., smaller numbers like 1 to 365).

Collisions.exe works best with larger numbers such as genetic code sequences, lotto combinations, social security numbers, etc. *Collisions.exe*, the sets-based option, is less accurate with small numbers such as birthday cases (e.g., inaccurate for birthdays of 200 persons in the room). *Collisions.exe* has a floating-point option that provides calculations with huge numbers. The procedure is accurate with an 18-digit precision. The probability beyond 18 digits is rounded up to 100%. That's the maximum of precision my compiler can trust at this time.

4. Relation of the Birthday Paradox Probability to Lottery

This is only a briefing of a much larger topic (will be presented in dedicated chapters of this book).

The relation of the Birthday Paradox formula to the pick-3 lottery goes further. The extrapolation was to use such a paradox to predict more easily the pick-3 lottery. It was suggested (in http://.www.rec.gambling.lottery Usenet) that a set of 16 pick-3 combinations would hit a boxed prize with a 50% probability.

The Birthday-paradox-pick-3 lotteries relation applies *strictly* to the *number of duplicate digits in a draw*. The probability of at least two common digits has absolutely no relation with the probability of *guessing two digits*.

Every pick-3 digit has the same probability to win. I generated all 1,000 straight sets in the pick-3 game, from 000 to 999. This Birthday Paradox program can generate all pick-3 and pick-4 sets, too. I used my specialized lottery software to do a statistical report (frequencies) for all 1,000 combinations. Each pick 3 digit shows a frequency equal to 300, regardless of position (boxed). Thus, the probability to predict one digit is p = 300 / 1000 = .3 (or 30%). There is no surprise there. The probability to predict two digits is exactly the product of the two individual probabilities: .3 * .3 = .09 = 9%. The probability to predict three digits is exactly the product of the three individual probabilities: .3 * .3 * .3 = .027 = 2.7%. That is, if we play *one* ticket. If we play 16 tickets, the probability grows to 43%. But again it has nothing to do with the *birthday paradox*.

The pick-3 game has a very specific form of playing *boxed*. In such play, 1-2-3 is the same as 2-1-3 and 3-2-1; 1-2-2 is equivalent to 2-1-2 or 2-2-1. The pick-3 game has 220 boxed possibilities. If we play 1 boxed ticket, the probability is 1 in 220 (0.45%). If we play 16 tickets, the probability is 16/220 or 7.27%. If we play all 220 boxed combinations, we are guaranteed to win!

• There are some interesting facts, however. The *birthday paradox* calculates that the probability to get the same pick-3 combination at least 2 times in 100 trials is 99.4%. More clearly: this is the probability of a pick-3 straight set to show a *repeat within 100 drawings*. I checked the most recent 200 draws in Pennsylvania State Lottery.

I can get the same results at any given moment in a lottery history. There are 34 occurrences of pick-3 straight combinations that are repeats from the previous 100 draws. There are no situations without repeats longer than 100 drawings!

Furthermore, the *birthday paradox* also shows that if 100 players play independently one ticket each, the probability is 99.4% that at least 2 tickets have the same exact combination! You can ask a lottery agent to collect any 100 discarded pick-3 lottery tickets. The chance is really high (99.4%) that at least 2 tickets show the exact same combination (in straight order)!

The birthday paradox applied to a lotto 6/49 game. If 10,000 players play independently one combination apiece, the probability is 97.2% that at least 2 tickets have the same exact combination. The probability of *duplication* is 99.999789% if 100,000 combinations are played independently. It does not imply that the jackpot combination will be necessarily a combination played more than once! Of course, sometimes that's the case—and the jackpot is shared.

Equivalently, the probability is 97.2% that at least 1 jackpot combination will be a repeat within 10,000 drawings (something like every 100 years). There are many cases for lotto games with lower odds (probabilities). For example, the jackpot combinations have repeated numerous times in 5-number games.

5. Relation of the Birthday Paradox Probability to Roulette

The connection between *birthday paradox* and the game of roulette is extraordinarily appealing to some gamblers. Again, there is foundation to it: the roulette numbers certainly repeat after a number of spins. But let's not confuse the repetition of the roulette numbers with the probability of predicting the next spin!

If I consider 10 roulette spins at a time, in 72.7% of the cases, at least 2 of the numbers will be repeats. So if I play the last 10 numbers, the chance is very good (almost 3 out of 4 cases) that 1 of the numbers will repeat next! Wow! That would bankrupt every casino on the planet in a few days! If the strategy would hold true, the probability would rise to 99.8% that the next spin will repeat a number from the last 20 spins! Virtually, play the last 20 numbers and win every time. The cost is 20 units, the payout is 36 units; therefore, that player would make a profit of 16 units in every play!

The cold truth is that the famous and appealing Birthday Paradox merely shows the percentage of sets with duplicate elements in the total elements of an exponential set. Devising a gambling strategy is a whole lot more complex issue. It will be dealt with in a future chapter.

I did check several roulette tables in Atlantic City. I did not find a roulette marquee showing unique numbers only! The marquee at every roulette table showed repeated numbers. Out of 15 numbers (on the electronic display), some were repeats—from 3 to 7 repeat numbers. Problem is the skips between Birthday Paradox situations reached 8 spins sometimes! Probably some players wait for 5 or 6 or so skips and then apply the Birthday Paradox. The average amount of unique numbers to play is 12. One must win in 2 spins in order to make a profit or apply a limited-step martingale bet.

6. The Birthday Paradox and the Social Security Number

The government of the United States of America assigns its legal residents an apparently unique number called *social security number (SSN)*. The number is in the following format as far as I know:

NNN-NN-NNNN

Theoretically, total possible numbers go from 000-00-0000 to 999-99-9999, or 1,000,000,000 (one billion).

Obviously, a system as such is nonfunctional in a nation like China or India. Each nation has over one billion people. When the concept of SSN was devised, one billion numbers seemed sufficient to cover the population of the United States for a long period.

But how about deaths? I hope they do not reassign social security numbers from the deceased to the living. I heard of crimes committed by using social security numbers of dead Americans. At one point, the one billion pool will not be sufficient anymore. The computers are increasingly powerful. They can handle more and more easily huge numbers. I think of a social security number in the format:

NNNN-NNN-NNNNNN

Such a number is 10,000 times greater; that is, it can cover up to 10 trillion people! The computers can handle such numbers now. The computers evolve so rapidly that they will be a thousandfold more powerful every decade. The numbers of the deceased can be reassigned no earlier than every million years!

The point here is to *avoid duplication* in relatively small groups. I receive quite a few questions related to mathematics and probability theory. Some inquirers promised that liberties, even lives, were in jeopardy! How can one company uniquely identify employees without using the entire SSN? The entire social security number is a very important privacy issue. The law penalizes the publication of the social security number of any individual. One method to respect privacy is to use only the last four digits of the SSN. Problem is duplication has a very high probability.

There are only 10,000 possibilities, from 0000 to 9999. What is the probability that at least 2 persons have the exact same last four SSN digits in a group of 100 individuals? We can use our great freeware *BirthdayParadox.exe* or *Collisions.exe*, with the following parameters:

~ Lower bound = 0
~ Upper bound = 9999
~ Total elements (persons) = 100

The probability that at least two persons have the same last four SSN digits is 39% or 1 in 2.6 (better than 1 in 3 individuals share the same last four SSN digits).

If we take into account the last five digits (0 to 99999), the probability goes way lower: 4.8% or 1 in 21 persons.

Using the new SSN format (I mean, my proposition), with the last category consisting of 6 digits (0 to 999999), the probability that at least 2 persons have the same last 6 SSN digits is 0.05% or 1 in 200. It is likely that a group of 199 persons will not have a duplication of the last 6 SSN digits.

I have been able to calculate the birthday paradox for the current format of the social security number. If the social security number would be assigned randomly, the repeats would be inevitable, even in relatively small samples. If 100,000 social security numbers were issued randomly, the birthday paradox probability would be 99.33% to get at least 1 duplication.

The probability grows to 99.9976% for 1,000,000 persons (very high degree of certainty)! If you try to calculate the birthday paradox probability for more than 1,000,000 persons, the results are no longer reliable. The probability goes down instead of climbing toward 100%. That's so because of the programming tricks I employed. Such tricks allow for birthday paradox calculations applicable to huge numbers. As soon as the probability starts declining, it's a sign the probability is virtually 100%. Run instead *Collisions.EXE*. *Collisions.exe* has a floating-point option that provides calculations with huge numbers. The procedure is accurate with an 18-digit precision. The probability beyond 18 digits is rounded up to 100%. But that does *not* mean absolute certainty—ever!

7. The Birthday Paradox and the Genetic Code, DNA, Forensic Coincidences

In the year of true grace 2000, the scientists have mapped the human genetic code (genome). That is, they mapped all genes and genetic sequences in DNA (deoxyribonucleic acid).

The super molecule, DNA, is present in every cell of living entities, from single-cell forms of life (the very beginning of the beginning of life) to *Homo sapiens* (we, the humans, representing the highest form of known life).

There are a few billion sequences of genes in the human genetic code—a gigantic number by normal human measure. However, that number is modest by numerous other standards. The experts in genetics, and especially the forensics, consider that duplication of the genetic codes is almost impossible. They might say that no two humans have the same identical genetic sequences. Actually, they use a cliché such as "The odds are like one in one billion . . . almost impossible!"

They, the modern forensics, appear to be overly religious (reminds me of a superintelligent mystic named Einstein). The birthday paradox proves otherwise. There is no god or a Super Universal Power to dispense of every genetic sequence in lexicographical order. Randomness is the highest (supreme) attribute of the Universe. All things collide (interact) randomly. Nothing comes in sequential order (lexicographic) as if ordered by an intelligent force. That's why we haven't found any perfect shapes in the Universe, except for highly successful human attempts (e.g., circles, spheres, pyramids, cones, etc.).

The Universe is more like a gigantic lottery: all things come and go randomly. Humans are no exception. The human DNA is no exception. Every human individual gets his/her genetic sequences (DNA) absolutely randomly. Randomness implies uniqueness and repetition. I believe I know all probability formulas that calculate the degree of uniqueness and the degree of repetition for all phenomena in the Universe. All we need is to calculate all possible events (N) and all favorable cases (n).

Let's consider a very generous number of gene sequences: 10 billion (10,000,000,000 in U.S. math). Let's suppose just 1 million human individuals. The probability is 99.9999% that at least 2 humans will have absolutely identical genetic codes! Let's take a smaller city of only 100,000 (one hundred thousand) inhabitants. The probability is 39% that at least 2 humans will have absolutely identical genetic codes! There is a better than *1 in 3* probability that a crime could have been committed by 1 of at least 2 persons in that 100,000-inhabitant city!

Always keep in mind the *at-least* emphasis—and never forget the "one in one billion" forensic statement in the court of law! Teach the forensics how to run that incredulously great and precise piece of software known as *Collisions.exe* to *Homo sapiens* around Terra. The forensics simply confuse chance of duplication for chance of *guessing* correctly two DNA sequences consecutively!

Keep also in mind that the genetic sequences are not generated uniquely by some cosmic supercomputer. *The Cosmic Supercomputer* would be a more suitable metaphor for *God* in the third millennium. The genetic sequences are generated randomly as the result of free interaction of forces in TheEverything. The Cosmic Supercomputer does not issue DNA (genetic sequences) based on the law or breaking the law.

Whose law after all? God's Law? If we look at all religious (sacrosanct) texts, the gods have been the most hideous of criminals! They have been the worst mass murderers imaginable! But the gods have gotten away with crime because they have no DNA! Anyone found Jupiter's DNA? Or Heracles'? If they had discovered the DNA of Aphrodite, all rich and famous women would kill to be Aphrodite-cloned!

I want to tell you something unsettling. I sent a dozen e-mails or so to the most reputable organizations in the genome field. I asked them a very simple question: *"How do you calculate the number of possible genetic sequences?"* They all took the vow to answer any question. They are funded to answer any question. Yet they have not answered my simple question for several years now. They never will it seems. I'm afraid their mathematics is real bad. They feel it unconsciously. My question was not meant to embarrass anybody. I wanted to help, if help was needed. I believe I am in the best position with calculating the total possible number of genetic sequences. If we know that, even closely approximate, we will know better the probabilities of duplication including for DNA sequences.

8. The Birthday Paradox and the Chance of Duplication of Intelligent Life in the Universe

We know beyond reasonable doubt that *intelligent life* does exist in the Universe. We can witness it right here on Earth, the third planet from the star we call *Sun*.

A big question has been asked incessantly on this planet since life entered the stage of *self-awareness*. Is there intelligent life in other places in these cosmic surroundings of Terra?

Lucky us, we are in the familiar territory of calculating the degree of certainty of repetition or duplication. We know for sure one element of the equation: intelligent life exists.

Is the element named intelligent life *unique*? Or is intelligent life always a *repeat*?

We have already learned that one element cannot exist *absolutely* while the opposite element is missing *absolutely.* There is always a *probabilistic balance* between two opposites. Randomness always strikes a balance.

We apply here the by-now-famous (very well-known, that is) ***Ion Saliu's Paradox of* N *Trials***. Let's say that *N* represents total possible cases the Universe can handle. It is even more fitting to work at a universal scale. The Universe is infinite; therefore, nothing could be more fitting but to consider that ***N* tends to infinity**. We know that human life exists. What is the repeat probability of human life, exactly as is, in other parts of the Universe when *N* tends to infinity? The ***Ion Saliu Paradox*** proves that the degree of uniqueness is, approximately, 63% while ***the degree of duplication (repeatability) is 37%***.

True, the *odds against a duplication* of human life, at any time, in any place of the Universe, are almost *2 to 1*. Nevertheless, a better than 1-in-3 chance of duplication looks real good to me. After all, you or I have a chance to see the very tomorrow with odds worse than winning a 5/37 lotto game! Read the **Probability Caveats** chapter. I am sure, with a high degree of certainty, you will—we will—be here again tomorrow! Thus, the probability of duplication of intelligent life is quite high in this Universe as it was in a previous Universe or will be in a future billions-of-billion-year Universe. It's mathematical.

Granted, we have **not** discovered a bit of evidence of intelligent life in places other than planet Earth. The most compelling test of intelligent life is the existence of ***perfect shapes***. Only humans, possessors of intelligence, are capable of creating perfect shapes. We might call them as well ***geometric shapes***, as humans alone comprehend geometry.

Perfect shapes such as lines, 2-D surfaces (circle, triangle, square, pentagon, hexagon, etc.), 3-D objects (pyramid, cone, cylinder, sphere, etc.).

I was asked this question in a public forum: *"What is the probability for a river to flow in absolute straight line all the way, without deviation?"*

The *perfect-line river* is a beauty of a mind game! There is absolutely no evidence of a river flowing in a perfectly straight line—nowhere here, at home, on planet Earth, and nowhere in the known Universe.

How about the probability of a mountain in the shape of a perfect pyramid? How about the probability of a tree in the shape of a perfect cylinder, in a perfectly vertical position? There is absolutely no evidence of such shapes on our planet. We can think of other perfect shapes or geometric shapes: circles and spheres. How about the probability of a lake or sea in the shape of a perfect circle? How about the probability of a rock in the shape of a perfect sphere?

On the other hand, I don't think the probability of one of such shapes is zero. For sure, such probability is equal to the probability of any other shape. The Universe is infinite; the matter takes an infinite number of shapes. Therefore, the probability of an individual shape is close to zero, but never equal to zero. Who knows, maybe somewhere in this Universe, there is a mountain in the shape of a perfect pyramid! If not in the present, maybe in the infinite future. The Universe is the result of absolutely free interaction of forces. Randomness is the supreme form of free interaction. Randomness creates all those shapes of matter. Randomness also creates all those lottery combinations where *1-2-3-4-5-6* is noticeably missing! The same individual probability, but apparently, Randomness Almighty prefers certain shapes or combination thereof.

Only humans (or any creature belonging to the intelligent life in the Universe) attempt to override randomness. Humans oppose randomness by using the rational tool named *rule*. The rule expects a strict result, not a random result. The rule, for example, is to plot all the points equally distant from a fixed point (the center). The strict result is the geometrical figure known as *circle*. Things are relative, however. There is no strict or absolutely certain result. A multitude of factors (all of them random) can prevent the human subject from carrying out a rule. Besides, the opposition to Randomness is ephemeral in the complexity of universal time.

The lotto combination *1-2-3-4-5-6* seems to be heavily ordered (compliant with a strong rule). The words of natural languages are also heavily ordered by the rules of grammar and syntax and pronunciation, etc. A program I wrote, ***Writer.exe***, will not generate any meaningful sentences in any language anytime soon. Almighty Randomness loves those random patterns of all forms and shapes!

Only the humans create perfect shapes in very short lengths of time. Wherever self-aware intelligence exists, perfect shapes exist as well. I wonder why the gods of this Universe create such irregular, rugged shapes. The Earth is a curved but irregularly shaped football with peaks and valleys. The entire Universe is curved, but it must be irregularly shaped and rugged as well. Looks like the gods don't possess intelligence! A figure of speech—the gods are supposed to be infinitely more intelligent than the mere mortals. The Universe itself and everything inside the Universe would be perfectly shaped—an infinity of geometric forms like Plato's Forms (or Ideas). It is Time to put the gods to rest. We know better now.

This is a very compelling proof of the inexistence of the gods, of any gods, or an Almighty God. It's a visual proof of the impossibility of the existence of the deity. The Deity—Ra, Zeus, Deus, Jehovah, Brahma, etc. many times—is perfect. Deity is conceived as Perfection itself. The deities are conceived perfectly intelligent. They are infinitely more intelligent than the mortal humans.

Well, then the Universe should have been perfectly shaped. Alas, O immortal ones, only these mere mortals, *Homo sapiens*, create an environment of geometric shapes. They, the humans, call the geometric shapes ***perfect shapes***. Perfect because they abide by very strict rules. Perfect line—absolutely straight. Perfect circle—absolutely round shape, all points being equally distant from a fixed point named center. The orbits of the celestial bodies only seem geometrical or perfect. They are far from being elliptical. Only we, the humans, approximate them as being elliptical. It's so because we fear that *nihil*, as in *Nihil sine dio* (*Nothing without God*). The shape is perfect; we cheat ourselves. Therefore, (a) God created it, but not before beating badly that bad guy, Nihil.

And then, O fearful one, there is the mathematical proof of the absurdity of the Deity concept. It is reflected in the Fundamental Formula of TheEverything (FFG). Mathematics demonstrates beyond reasonable doubt that absolute certainty is absolute absurdity. Only irrationality hides that rational demonstration. Not to be denied, irrationality has proven to be one of the most trusted shields for the rational creature, *Homo sapiens*. Can Deity (or God) be perceived outside Absolute Certainty? Who would believe in a god that's even slightly uncertain? Nobody! Just an infinitely small amount of uncertainty would render God just another force in the Universe. God would be just another powerful man, like a king, nevertheless a man, therefore limited in time and space.

Mathematics always provides the most compelling proof of everything—or invalidation of everything.

Seeing is believing, they say. Our *eyes* currently are not powerful enough to look for perfect pyramids or perfect spheres in Cosmos. But we are smart enough to employ other methods of exploring for intelligent life elsewhere. The radio waves, for example, can indicate nonrandom patterns, therefore the result of humanlike intelligent activity.

Chapter VI

IMPORTANT PROBABILITY CAVEATS

1. The Fundamental Mathematical Caveat: *Probability* Equals *Degree of Certainty*

There are many confusions or caveats in probability theory. Most notably, most people confuse the individual probability, ***p***, for degree of certainty, ***DC***, and ignore the number of trials, *N*.

That is the innocent part. There is a minor but very vocal group of humans who create the confusion intentionally. Their intention is hate based and serves selfish purposes. I can tell you firsthand that some react with intense hostility because they want me to remove such writings from my Web site and then they could take my ideas and get credit for them! The casinos also have a big problem with my theories because they show the reality. Such mathematical truth can lead to superior play from many gamblers. The casinos do not want people to realize the distinction between *probability* and *degree of certainty*.

This issue has caused the most trouble for me in public places such as newsgroups or forums. I trigger intense hostility toward yours truly as soon as I make the distinction between the three elements. The probability, ***p***, is an abstract concept. It is the constant of a particular phenomenon. It has no life of its own. If you conduct one trial for an event of probability, ***p***, then the degree of certainty, ***DC,*** is equal to ***p***.

It is the only case when $DC = p$. That is the main reason of the fundamental confusion that started this chapter. If we toss a coin 1 time, the degree of certainty to get heads is .5, which is also the probability of tossing heads. But the degree of certainty is equal to .75 to get heads in 2 consecutive tosses. So if the first toss was tails, I expect heads with a degree of certainty of 75%.

That's when the roar of hatred reaches a peak! Many will shout that the expectation is still 50% before the second toss! They only see the constant probability, *p*, but are blind to the other two elements: number of trials and degree of certainty!

> *1) The probability against your winning the lottery is non-zero.*
> *2) Therefore the event against your winning the lottery will repeat ad infinitum.*
> *3) If you bet on heads in coin-tossing, then the tails will repeat an infinite number of times.*

One can write books against the previous three "points." I can afford to only touch those points. All humans of Reason will easily comprehend and validate my points.

Everything in the Universe has a nonzero probability, except for absurdities. Absurdity is not to be confused with impossibility. Mathematics defines impossibility as an event of a probability infinitesimally close to zero. Absurdity is a phenomenon without a valid mathematical relation (formula, equation, etc.).

One common caveat is absolute certainty. Most humans believe that certain events are certain while other phenomena are random. A human considers his/her tomorrow as certainty. The same human considers his/her 5/39 lotto game to be random. Well, it isn't just like that, that simplistic!

I will make use here of very conservative statistics regarding unexpected loss of life in the United States of America. It is a very conservative statistic that at least 50,000 Americans die *unexpectedly, therefore randomly*, of gunshots every year. At least 50,000 Americans die *unexpectedly, therefore randomly*, in traffic accidents every year. At least 150,000 Americans die *unexpectedly, therefore randomly*, because of taking prescription drugs every year. Now that amounts to 250,000 Americans who had been certain of their tomorrows. Unfortunately, they died with the wrong perception of randomness and certainty.

Let's take the above stats to a daily basis. Let's further the calculations: 250,000 / 365 = 685 Americans who die unexpectedly every day. We can calculate here a parameter named randomness of *ungodly death* by, say, a notable statistician in the fifteenth century. There are around 300,000,000 (three hundred million) strong Americans.

685 / 300,000,000 = 0.00000228333 . . .

The odds of dying unexpectedly (randomly) in America is *1 in 437,956*. Wow! It's kind of scary! It's somehow easier to win the 5/39 lotto jackpot than to survive to your next day if you live in America! Unfortunately, the situation is even worse, much worse in most places of this great world of ours!

By the way, the odds of winning the 5/37 lotto jackpot is *1 in 435,897*. So the odds that you will die tomorrow without notice are about the same compared with winning the 5/37 lotto jackpot with one ticket in one try!

Does it imply that the humans should not take any reasonable action because the probability of their dying randomly is nonzero? Of course not! The reasonable humans take a multitude of actions with the expectation to improve their odds of surviving each and every day.

The probability of the moon colliding with the earth is greater than zero. In fact, the moon was born as a result of a collision of the planet earth with a planet the size of Mars. Now shouldn't reasonable humans take any action in order to survive? Should they just cover their heads in ashes and wait for the collision of the moon with the planet Terra? Of course not!

2. Events Cannot Repeat *Ad Infinitum*

How about the infinite repeatability of adverse phenomena? They say, well, you bet on *heads* in tossing a coin. Guess what? The *tails* come up against your desire! You say, well, now my chance of winning is higher: the *heads* will be more likely to come out. Horror! The *tails* come out again! Yes. It happens quite frequently—it is not against mathematics. But the absurdity starts with vociferous Psychosamas claiming that your adverse phenomenon—*tails*, in this example—will come out again and again and again *ad infinitum*—an infinite number of the same outcome.

As we already have seen, the theory of probability is based on a fundamental formula. The probability, p, is defined as *the rapport of favorable cases,* n, *over total possible cases,* N: $p = n / N$.

That parameter, *p*, is a constant for its respective phenomenon. For example, *p* is always ½ (or 0.5) in coin tossing; 1/6 (or 0.16666667) in getting one particular face of a die; 1/1000 (or 0.001) in hitting one particular pick-3 lottery number (e.g., 0-1-3). The parameter, *p*, remains the same in every calculation we do in probability theory.

If I take 10 coin tosses, I always set p = 0.5 in toss no. 1, toss no. 2, toss no. 3, . . . toss no. 10. I know that the probability to get *heads* is 1 in 2 in the first toss, 1 in 2 in the second toss, 1 in 2 in the third toss, . . . 1 in 2 in the tenth toss. Yet the probability to get *10 heads in 10 consecutive trials* is .5 ^ 10 = .000976. You do the experiment. You'll definitely notice that 10 consecutive heads occur about once in 1,000 experiments (*one experiment* here is tossing the coin 10 times).

You take the pick-3 lottery game. There are 1000 possible outcomes, from 0-0-0 to 9-9-9. The individual probability, *p*, is 1/1000 (or .001)—always. The probability to win with the pick-3 combination 0-1-3 is .001. Let's suppose that 0-1-3 hit today. What is the probability that 0-1-3 will hit again the very next drawing? There is a subtle distinction here.

You can run my free software ***Collisions.EXE***. The extraordinarily tiny program can perform huge-number calculations with amazing ease. Don't they say that great things come in small sizes? This pick-3 example represents a case of probability of coincidences or probability of collisions as we learned in a previous chapter. The most famous case in this category is represented by the so-called *birthday paradox*.

Invite, randomly, 23 persons to a common place, a room, for example, or a desert tent. You can pick 23 camels for that matter; the only requirement is to know precisely the birthdays of every camel. The birthday paradox tells you that the probability is 50-50 that at least two camels (or horses or any creature with a pedigree) share the same birthday if 23 individuals are concerned.

Many people are victims of confusion here. They confuse the *probability of the birthday paradox* with the *probability of guessing the birthdays of two individuals*. I made the same confusion in the past. The most important lesson is to learn from our mistakes. And the best method is to find the right path to mathematics.

Okay, the probability of a pick 3 repeat is 0.10000000000000002%. That's about a little better than .001. So if a pick-3 combination hit today, it has virtually the same chance to hit the very next drawing.

We must make certain that the following point is understood. The probability above refers to *any* of the 1,000 pick-3 combinations. We do not nominate *one* particular pick-3 combination to repeat the very next draw.

If 10,000 lotto 6/49 drawings are conducted, the probability is 97.2% that at least two combinations will be the same (duplicates). If 5,000 lotto 6/49 drawings are conducted, the probability is 59.1% that at least two combinations are the same. If 1,000 lotto 6/49 drawings are conducted, the probability is 3.5% that at least two combinations are the same. If 100 lotto 6/49 drawings are conducted, the probability is 0.0035% that at least two combinations are the same! You can see it takes a large number of events to have a lotto 6/49 repeat combination. How far is that from a combination repeating in each and every lotto drawing?

When *precisely* one particular combination will repeat, the probability is significantly lower. If we take, for example, the *exact* pick-3 number 0-1-3 to repeat in the next draw, then the probability may show *1 in one million*. That was the subtle difference I was talking about a while ago.

There is so much more to it! Yes, the probability of repeating *any* outcome the next drawing is virtually the same compared to coming out just before that. Delving deeper always harvests more pearls. I analyzed—thoroughly, I believe—real-life roulette spins recorded at the Hamburg, Germany, casino (Hamburg Spielbank).

The software (my own creation, ***SuperRoulette.exe***) shows that the roulette numbers do not like to be followed by themselves! Like it or not, 25% of the roulette numbers repeat more than 55% of the time after a particular roulette number. That is, any particular roulette number always prefers other numbers to pair with other than itself. Well, I know why, and you can too. It's about ***FFG***. Again, another big book can be filled around this topic. Suffice to say for now that the numbers are followed by different numbers significantly more often than being followed by themselves. The length of data I analyzed is statistically significant: about 10,000 spins. That's more than 250 times total possible outcomes in the double-zero roulette. And that is *real*-life data.

It is the same thing when considering the probability of a pick-3 combination to match its date expressed as a 3-digit number. Again, my software comes to the rescue: ***DatePick3.exe***. The tiny program calculates the probability for the *any* cases (*any* as opposed to *exactly one particular case*). If we consider *any* of the pick-3 combination matching its corresponding 3-digit date format, then the probability is *1 in 1,337* (not *1 in 1,000*).

If you check all pick-3 drawings in the world, you'll find that the results are very close to 1/1,337. Now if you take one particular pick-3 combination (e.g., *1-1-1*) *to hit on January*

11, expressed as 1/11 or 11/1 and check all the drawings in history, the probability will show 1/365,000. I know there are not enough real-life drawings to accommodate precise calculations. Still ***DatePick3.exe*** will serve you excellent simulations.

Oh no, please! Don't take the *9-1-1 on September 11 (9/11), 2002, in New York lottery* example. It was a farce founded on humane reasons. I still believe that the New York lottery officials must come forward with the **truth**. What they did was understandable. I have no problem with that. I only have a serious problem with hiding the truth. It would be highly entertaining to read about all the discussion behind closed doors. I'm sure that the governor and the mayor were involved, and they pushed the green-light buttons.

3. The Probability Theory and the Mechanical Elements

There is another caveat regarding probability. *Is the Fundamental Formula of Gambling still valid IF we change the coins from one toss to the next?* Are the probability formulas still true when the lottery commissions change the ball sets of the drawing machine?

The answer is ***yes***. The probability formulas have nothing to do with the mechanics. The ***p*** and ***n*** and ***N*** and ***DC*** are abstracts. They are numbers. The probability to get *heads* is 1/2 or .5. The *p* doesn't care what coin was tossed. The denomination or the nationality of the coin doesn't matter at all. The only requirement is that the coin is unbiased. In fact, the probability would love to have a new coin every time for every toss. Thus, the unbiased requirement is satisfied. A coin will get biased after a number of tosses (yes, the material, the metal, is important!).

One more important requirement is the ***serial*** factor. The experiments (trials, coin tosses, etc.) must be serial. They come ***one after another***. The probability formulas take into account one serial event at a time. The opposite is ***parallel*** events. For example, two coins are tossed at the same time. The events are parallel. The only way we can research them is to record every coin separately. You conduct an absurdity if you record the spins of several roulette wheels as a single phenomenon.

Big books can be filled on these topics. Randomness and Certainty are fighting brutal wars every split of a second of every second and minute and hour and day and week and terrestrial year and century and millennium and era and galactic era and universal era *ad infinitum*. Each and every one of us might die tomorrow—absolutely unexpectedly and randomly. But should we give up all of our strategies and systems of living because the unfavorable odds are nonzero? I didn't think so.

Yet filling big books with all kinds of formulas and strategies can have consequences. I learned that I am on the *blacklist* of every casino in the United States! I am blacklisted because I am supposed to possess every winning formula for every type of gambling! I swear by the dog in Egypt, I bear it as a badge of honor, my being blacklisted for the sole motive of my knowledge by all the casinos in the great nation of USA!

4. The Past Is One of the Essential Elements of Probability Theory

The role of the past is the birth certificate of probability theory. We already talked about the case known in history as *de Méré case*. The dispute was brought to the attention of a wise man of the era: the philosopher and mathematician *Blaise Pascal*. Two gentlemen played a best-of-three backgammon match. The game had to be interrupted after the first game. Thus, one of the gentlemen had a 1–0 lead. The two gentlemen agreed to resume the game as soon as the circumstances permitted. The gentleman in the lead wanted to continue with game 2. The competitor, who was trailing, wanted a fresh start. Restart at 0–0 that is.

"A fresh start isn't fair," Blaise Pascal might have expressed in his impeccable French. The player in the lead had already an advantage that he had earned. Pascal was new to the field of mathematics known as theory of probability. As a matter of fact, Pascal was inventing probability theory! So his calculations looked rather complicated but nonetheless convincing. The past counts.

Today we possess many more tools, including a more-advanced science, plus we have computers and software. We will use here my free program ***SuperFormula.exe*** to perform probability calculations no matter how complicated. The *de Méré* problem can be solved by the *probability of the binomial distribution*. It is also known as the *probability of exactly* M *successes in* N *trials*.

We must suppose that the two players are equally skilled; thus, the winning probability is 1/2 or .5. The first player to win 2 games wins the match. The trailing player must win the next 2 of 2 games. The binomial distribution probability for 2 successes in 2 trials is .25 or 1 in 4. The player with the lead needs to win only 1 of 2 games. The binomial distribution probability for at least 1 success in 2 trials is .75 or 3 in 4. Clearly, the player in the lead has a better chance to win the match after its resumption.

The second method is even more illustrative and simpler to follow. We will use the *exponential sets* or *Ion Saliu sets* (as presented in the combinatorics chapter). There are two players, *1* and *2*. They play three games. Total possible outcomes: $2^3 = 2 \times 2 \times 2 = 8$. They are

1, 1, 1 (that is, player no. 1 wins all three games)
1, 1, 2
1, 2, 1
1, 2, 2
2, 1, 1 (player no. 2 wins the first game, loses last two)
2, 1, 2
2, 2, 1
2, 2, 2 (player no. 2 wins all three games)

If player no. 1 won the first game, then only four outcomes are valid:

1, 1, 1
1, 1, 2
1, 2, 1
1, 2, 2

Three of the four lead to the final success for player no. 1. Only one of the four cases is favorable to player no. 2: 1, 2, 2.

Thus player no. 1 has a *3/4 = 75% chance* to win the match after resuming with game no. 2.

What if the bet were the *best of five*? Again, the match resumes after one player leads *1–0*. The first player to win 3 games wins the match. The trailing player must win the next 3 of 4 games. The binomial distribution probability for at least 3 successes in 4 trials is .31 or 1 in 3. The player with the lead needs to win only 2 of 4 games. The binomial distribution probability for at least 2 successes in 4 trials is 0.69 or 1 in

1.5. Clearly, the player in the lead has a better chance to win the match after its continuation.

The American professional sports are founded on the *play-off* concept. The *best of seven* is the predominant format. The team with the *first four wins* advances or, ultimately, wins the championship. Let's analyze three situations.

After 1–0
- the trailing team needs to win at least 4 of the next 6 games; probability is 34%
- the leading team needs to win at least 3 of the next 6 games; probability is *66%*

After 2–0
- the trailing team needs to win at least 4 of the next 5 games; probability is 19%
- the leading team needs to win at least 2 of the next 5 games; probability is *81%*

After 3–0
- the trailing team needs to win exactly 4 of the next 4 games; probability is 6%
- the leading team needs to win at least 1 of the next 4 games; probability is *94%*

Of course, a professional game is not exactly a coin toss. The teams are pretty well matched nevertheless. They call it parity. It was believed to be impossible to come back from a 3–0 deficit and win a series. I think it happened only once. The Boston Red Sox defeated the New York Yankees in 7 games after trailing 3 games to none. The Red Sox went on to win the pro baseball championship (the World Series) in 2004. If the games were played on neutral sites, the statistical data would be even closer to the theoretical probabilities.

Again, one easy method is to apply the *Saliusian sets (Ion Saliu sets)*. The number of elements in the set is calculated by exponentiation (raising to power). My software easily generates all possible sets.

There are problems like this one. Suppose you are told that *two visitors will show up at your place*. Each visitor can be male (M) or female (F). You are told that *one of the visitors is female*. What is the probability that the second visitor is also female?

There are 4 possible cases (2 to the power of 2):

FF
FM
MF
MM

However, one case is invalid: MM. We already know that one of the visitors is F. In only one of the three cases, both visitors are females: FF. Thus, the probability that the other visitor is also female is *1 in 3* (*.33*). The probability is *2 in 3 (.67)* that a *female* is accompanied by a *male* visitor.

This problem can have a twist however. You are told that precisely the visitor to the left is female. What is the probability that the visitor on the right-hand side is also a female? The only valid cases now are FF and FM. The probability that the visitor to the right is also female is *1 in 2 (.5)*.

To see more clearly the role of the past, we can toss two coins. We know that one of the tosses was *heads*. What is the probability that the other coin showed *heads* also? Same as above.

I have been aware of the role of history (past) in probability for a long time. The events can repeat—and they will repeat.

Also importantly, the events will miss—and they will miss or **skip**. The skips play an essential role in my theory, systems, and software.

5. How About the Streaks? Probability Is the Theory of Streaks

I looked at events from two angles:

~ Events cannot repeat ad infinitum (during a lifetime anyway).
~ History plays an important role in determining future events.

We can deduce a third important element in probability: *streak*. "It's all about the streak!" some gamblers say. "I am in a hot streak." A baseball hitter is said to be in a slump when he goes through a long losing streak (not hitting the baseball).

The casinos, especially online gambling sites, want players to believe that losing 10 hands in a row or 20 or even 100 hands in a row is a natural fact of life! But guess who is losing? Yup! Only the player! Don't ask why the house doesn't lose 20 or 100 hands in a row! You might receive really bad phone calls from unknown numbers, like I do now and then!

We usually divide the streaks in two categories: *winning streaks* and *losing streaks*. In reality, they represent one and the same concept. An event occurs or happens or appears or hits in a number of consecutive wins. We call them streaks. The streaks have a length (i.e., number of consecutive hits). The length can be from 1 to *N*. We put Infinity aside for it won't happen in a lifetime. In truth, Infinity doesn't occur at any given moment—TheEverything is finite at any sharp moment in Time.

I toss a coin. I am interested in *heads*, H (just for the sake of example, not because I am a heads-up guy!). I can record one *heads* followed by *tails* (T). The streak is 1 in this case. I can record also 1 *tails*, then 2 *heads* followed by *tails*; the streak is 2 in this case. We can symbolize the case as THHT. A THHHHHT case represents a five-streak for *heads*.

Do the streaks occur equally regardless of length? Of course not. Only the above gambling entities want gamblers to believe that. Mathematics doesn't agree, however. The shorter the streak, the more often it occurs in a given number of trials.

You can easily verify by tossing the coin and recording the results. If you do it 1,000 times or so, you will notice that the streak no. 1 occurred about the same number of times as the other streaks combined. You will also notice that the 2-streak is twice more frequent than the 3-streak and so on. The same is true for tails since it has a probability equal to the probability of heads: 1 in 2 (or 1/2 or .5). The ratio between various lengths of streaks is determined by the probability of the event.

Again, I already wrote the software for you. It calculates very easily how many streaks of a particular length you can expect in any given number of trials. The name of that probability program is ***Streaks.exe***. The probability program belongs in the software category 5.6 (scientific software).

How many streaks of 8 *heads* in a row? There is 1 (one) such a streak in 1,000 tosses. The same is true considering 9 *heads* in a row in 1,000 tosses. But the program calculates 0 streaks of 10 consecutive *heads* in 1,000 tosses. That is, the result is closer to 0 than to 1. Nonetheless, it is possible to get 10 consecutive *heads*. The Fundamental Formula of Gambling calculates what the degree of certainty is to get 10 consecutive *heads* in a number of trials. Just run the aforementioned software of mine, ***SuperFormula.exe***.

Chapter VII

MORE PROBABILITY PARADOXES, FAMOUS PROBLEMS

1. Introduction to the Concept of Paradox

Encyclopædia Britannica defines the **paradox** as "apparently self-contradictory statement, the underlying meaning of which is revealed only by careful scrutiny." We already learned about the Birthday Paradox. Most people find it paradoxical that in a group of 23 randomly chosen persons, at least 2 of them share the same birthday with a fifty-fifty chance. The impression of paradoxical: most people confuse the chance of duplicate birthday with the probability of guessing 2 birthdays in consecutive tries!

There is also a famous paradox from ancient times: *Zeno's Paradox.* This is a philosophical paradox not related to theory of probability. Zeno of Elea, a Sophist, was determined to prove that *motion* was a logical impossibility. Zeno argued that it was impossible for Achilles to surpass the tortoise in a footrace!

I was the first one to solve *Zeno's Paradox of Achilles and the Tortoise*. I wrote one dedicated essay at SALIU.COM. Zeno's construct is not even a paradox—it is an *absurdity*. In Zeno's false argument, the objects reach various positions in *space* in *no time*. There is no Time in Zeno's philosophy. If the objects reach various positions in *space* in *no time*, then all objects move at the same speed, equal to infinity! The absurdity is created by the separation of Space and Time. In truth, Space and Time are inseparable; they are fundamental attributes of the Universe.

I quote further from Encyclopædia Britannica: "The purpose of a paradox is to arrest attention and provoke fresh thought."

To that end, a paradox leads to a new way of looking at a *problem*. We read at the very beginning of this book that *"the most important questions of life are, for the most part, really only problems of probability."*

Thusly, we will deal here with *probability paradoxes* as *probability problems* that didn't seem to have logical solutions. Or the solution of a problem seemed to have raised eyebrows.

2. The Monty Hall Paradox: A TV Game Show

The name of this paradox comes from a TV game show popular in the United States. The host was named Monte Hall or Monty Hall. The name of the game show was *Let's Make a Deal*. The paradox is also known as the *choice of three doors*.

The host offers the player the opportunity to win what is behind one of three doors. Typically there was a really nice prize (i.e., a car) behind one of the doors and a not-so-nice prize (i.e., a goat; don't like goat or mutton) behind the other two. After selecting a door, Monty would then proceed to open one of the doors you didn't select. It is important to note here that Monty would *not* open the door that concealed the car. Thus, the host always *eliminated* one of the losing cases. At this point, he would then ask you if you wanted to switch to the other door before revealing what you had won.

Thusly, you're given the choice of three doors: behind one door is a car; behind the others, goats. You pick a door, say no. 1; and the host, who knows what's behind the other doors, opens another door, say no. 3, which has a goat. He then says to you, "*Do you want to pick door no. 2?*" Is it to your advantage to take the switch?

There is *no* real paradox here. Just generate all possible cases, and the problem becomes crystal clear. There are three elements in the problem. Total number of permutations of 3 is 3! = 1 * 2 * 3 = 6. The easiest way is to nominate the three cases as *1*, *2*, *3*. Nomination 1 represents the prize (i.e., the car); *2* and *3* represent the losing situations (i.e., the goats). Run my free software *PermuteCombine.exe*, option *Permutations*, then *Numbers Sequentially*. The program generates all 6 permutations of 3 numbers. I wrote another program that generates the permutations and also evaluates the Win/Loss (W/L) situations. Then the favorable cases are counted. Allow me, O wise reader, to present you with fragments of the output files. For every output is the child process of an input, and they both show clearly the picture of an entity.

The original game: N = 3 (three doors or choices)

```
Stay Change
==== ======
1 2 3 W L
1 3 2 W L
2 1 3 L W
2 3 1 L W
3 1 2 L W
3 2 1 L W
```

The digit in position no. 1 represents the player's choice. The player always selects first. Then, the host eliminates a losing possibility. The player is greatly assisted by having to make one of two choices. That's what makes many people believe that the player only faces a p = 1/2 choice. But they ignore what the total number of possibilities is and what the number of favorable cases is.

There is no doubt that the total number of possibilities is 6 or 3! (3 factorial or factorial of 3). No matter what, the player has always 2 *favorable* situations, those starting with digit **1**. It doesn't matter what door number that is. Digit **1** in position no. 1 means that the player guessed correctly the door number where the prize (car) hides. This situation occurs 2 out of 6 times, or *2/6 = 1/3*.

Of course, if the player stays (sticks) with her choice, the player wins. In other words, the player loses if she changes (switches) her first selection.

In 4 out of 6 cases, the digit in the first position is *not* 1; therefore, the player makes a losing selection at the beginning of the game. The complementary situation is *change* the selection. In other words, the player *wins* if she changes (switches) her first selection. The winning situation occurs 4 out of 6 times, or *4 / 6 = 2 / 3*.

This is a no-brainer really. The player is better served if changing his first choice. This was a TV game, not a casino game. No casino in the world would offer such a favorable-to-the-player game. If allowed to switch, I would bet on blackjack Dealer's hand on most occasions.

We can generalize this game. The probability to win in *stay* situations is *(3-1)! / 3!* For ***N*** choices, the probability is

P = (N – 1)! / N! = 1 / N

The probability to win in *change* situations is the complementary probability:

PC = {1 – [(N – 1)! / N!] = {1 – (1 / N)}

If ***N*** tends to infinity, the probability to win in *stay* cases tends to zero; the probability to win in *change* cases tends to 1 (certainty).

Let's analyze the *N = 4* case in our quest to find the general solution. In this case, there are four doors. The prize is behind one of the doors. The other three doors represent a losing choice. Again, ***1*** represents the prize; the digit in the first position represents the player's first selection. Here is what my program generated.

N = 4; four doors (choices); one prize (favorable case)

Stay Change
==== ======
1 2 3 4 W LL
1 2 4 3 W LL
1 3 2 4 W LL
1 3 4 2 W LL
1 4 2 3 W LL
1 4 3 2 W LL

Stay Again Change Again
========= ===========
2 1 3 4 L WL W L
2 1 4 3 L WL W L
2 3 1 4 L WL L W
2 3 4 1 L WL L W

2 4 1 3 L WL L W
2 4 3 1 L WL L W
3 1 2 4 L LW W L
3 1 4 2 L LW W L
3 2 1 4 L WL L W
3 2 4 1 L WL L W
3 4 1 2 L WL L W
3 4 2 1 L WL L W
4 1 2 3 L LW W L
4 1 3 2 L LW W L
4 2 1 3 L WL L W
4 2 3 1 L WL L W
4 3 1 2 L WL L W
4 3 2 1 L WL L W

The favorable cases are W—a total of 6 wins; that is, 3! = 6. Number of cases that qualify as *change again* is 18; that is, 4! – 3! = 18.

Of course, this situation is more complex. The host can choose to eliminate all the losing situations but one—after the player makes his first selection. In such a case, we are exactly as in N = 3. The probability to win *if stay* is 1/N; the probability to win *if change* is {1 – (1/N)}.

If the host eliminates only one losing door, the player has more nested choices to be made. That's why the program analyzes also "Stay Again" and "Change Again" situations. Still, the winning probability for *stay* is always 6/18 = 1/3. It is the same probability as in the original problem with 3 doors (choices). No-brainer: the player should always change. The winning probability for the *change* situation is always 1 – 1/3 = 2/3. If N > 4, we record the same situation for change and change and change, etc. The probability to win—*if stay*—is always 3! / (4! – 3!) = 6/18 = 1/3.

3. The Classical Occupancy Problem or the Coupon Collector's Problem

We have 10 balls numbered 0 to 9 and place them in a jar. We keep drawing one ball at a time and return each ball before the next drawing. We can ask a question like this: *How many trials will it take for all 10 digits to appear?*

The short answer seems illogical: we might draw one ball at a time *ad infinitum* without drawing every ball! Our Fundamental Formula of Gambling (FFG) proves undeniably that there is no absolute certainty!

This *paradox* is an old and apparently difficult problem. I searched on the Internet, and I also found references in old books. The honor of presenting the problem for the first time goes to the French mathematician Pierre Simon de Laplace in 1812. There have been several attempts to offer the most precise method of calculating the probability of **N *elements to appear with a degree of certainty***. I believe every method is an approximation.

I found some interesting new facts. It is undoubtedly accurate to say that a digit (i.e., from 0 to 9) will appear within 44 trials with a degree of certainty of 99%. But that calculation refers to just one of the digits. There will be situations when more than one digit will skip more than 44 trials (drawings).

Once again, I wrote software to simulate the Classical Occupancy Problem combined with Ion Saliu's Paradox: OccupancySaliuParadox.exe.

The program simulates a wide range of games from coin tossing, dice throwing, 10 digits, 3-digit combinations of the pick-3 lottery, etc. The user can select how many times to run the program. For example, the 10-digit case, drawing one element at a time, from 0 to 9. The user can run the program 10,000 times. I did it many times with just seconds per run. If the output is displayed on screen, the user can see also the individual elements generated randomly. When all 10 digits have been generated, the program ends the current run and starts the next. If the output is directed to disk, the execution is much faster and more comprehensive. The program saves to file total trials for each run. The runs are sorted in ascending order to calculate the median. Let's look at two sample runs.

10 digits

lower bound = 0
upper bound = 9
total runs: 10,000

Minimum number of trials: 10
Maximum number of trials: 164
Median number of trials: 30

one die

lower bound = 1
upper bound = 6
total runs: 10,000

Minimum number of trials: 6
Maximum number of trials: 91
Median number of trials: 15

You can run for the pick-3 lottery game: lower bound = 0; upper bound = 999. It takes much longer to generate all 1,000 pick-3 straight combinations. I saw runs of over 9,000 draws (trials) before all combinations came out!

The *median number of trials* is a useful parameter. It represents the number of trials for a degree of certainty equal to 50%. *SuperFormula.EXE* has a special function: *C = Calculate probability from median.*

The number of trials in the output files generated by *OccupancySaliuParadox.exe* are, in fact, skips. The digits *skip* various numbers of draws before *all* of them appear. In the die-throwing case, such probability for a median equal to 15 is p = 1/23. In the 10-digit case, such probability for a median equal to 30 is p = 1/44. These values get closer and closer to the *reverse of* N calculated by the Fundamental Formula of Gambling. It's option *F (FFG)* in *SuperFormula.EXE* that calculates *N* as

N = log(1 – DC) / log(1 – p).

It appears that the probability of the ***occupancy problem*** is approximated by the formula 1/N:

poc = log(1 – p) / log(1 – DC)
for a DC = 0.99 (99%)

That makes the ***Classical Occupancy Problem*** almost *mission impossible* for lotto games. How many drawings will it take to generate *all* combinations in a *lotto 6-from-49* game? If we set the degree of certainty to just 99%, it might take trillions of trillions of quadrillions of trials (lotto draws of 6 numbers each).

The fast computers of the today and the near future make possible simulations for longer and longer runs of *OccupancySaliuParadox.exe*. Therefore, much more precise relations will be found. They didn't have computers in 1812. Lucky us!

4. Ion Saliu's Paradox and the Law of Large Numbers

This paradox has been talked about here quite a bit. It isn't bias—it's necessity mind you! We analyze in this chapter the relation between probability and the famous ***law of large numbers.***

We just saw, with regard to the ***classical occupancy***, that it might take a huge number of trials for *all* the elements in a set to appear with a high degree of certainty. ***Ion Saliu's Paradox*** deals with the appearance probability of just *one* element of the set. Evidently, the degree of certainty required for *any one* element of a set to appear is lower.

The limit of the *degree of certainty DC* is **{1 - (1/e)}** when N *tends to infinity* for an event of probability $p = 1/N$ and a *number of trials equal to* N. The variable ***e*** represents the base of the natural logarithm and equals approximately *2.718281828 . . .*
The limit **{1 - (1/e)}** is approximately ***0.63212055 . . .***

They use the expression: *"In the long run, all things will appear with an equal chance."* How long is "in the long run"? Or how big is the law of *large numbers*? *Bernoulli* was the first mathematician to tackle such a problem (1713). His solution was a loose approximation. *Chebychev* came up with a tighter approximation known as *Chebychev inequality*—still a quite-loose approximation!

Ion Saliu Paradox of N *Trials* makes it easy and clear. First of all, *law of large numbers* is a subjective matter. There is no formula that can establish such a law. The Fundamental Formula of Gambling (FFG), however, gives more reasonable methods to establish *long run* or *large numbers* in correlation to a *degree of certainty (DC)*. If a person is satisfied with a degree of certainty equal to 99%, then the person will have a *personal setting of large numbers*.

Let's repeat the number of trials in *M* multiples of *N* (e.g., play 1 roulette number in 10 series of 38 numbers each). The formula becomes

$$1 - DC = (1 - 1/N)^{NM} = \{(1 - 1/N)^N\}^M = (1/e)^M$$

Therefore, the degree of certainty becomes

$$DC = 1 - (1/e)^M$$

If *M* tends to infinity, $(1/e)^M$ tends to zero; therefore, the degree of certainty tends to 1 (certainty yes, but not in a philosophical sense).

Actually, relatively low values of *M* make the degree of certainty very, very nearly 100%. For example, if M = 20, DC = 99.9999992%. If M = 50, the PCs of the day calculate DC = 100%. Of course, they can't approximate more than 18 decimal positions! Let's say we want to know how long it will take for all pick-3 lottery combinations to come out. The computers say that all 1,000 pick-3 combinations will come out within 50,000 lottery drawings with a degree of certainty extremely close to 100% (with today's computers).

How large the *large numbers* are depends on you. How high do you want the *degree of certainty* to be? That is the question. And the answer will be so easy and precise because we have computers today! This makes us more precise than Bernoulli or Chebychev or de Moivre—or anyone else who lived before us for that matter. They call it *evolution at work*.

5. Montmort's Problem or the Couple-Swapping Paradox

This is the original *Problem of No Coincidences*, known also by the French term of *Rencontres*. It was first analyzed by the mathematician *Montmort* (1708). It is also known as the *problem of not choosing one's hat*. I have been posed this type of problem in newsgroups and cyberforums. Evidently, humans always adapt older problems to their own times and contemporary concerns!

> *A number of married couples attend a wife-swapping party, with the pairing-off being completely random. What is the limiting probability that no man will be paired up with his wife, as the number of couples tends to infinity?*

Although the problem might be of more ardent interest to the anthropologists of the global village Internet, there is also profound mathematics involved. The solution necessitates combinatorics and the probability of the hypergeometric distribution.

Let's start with the *combinatorics*, specifically with just 3 couples—3 men and 3 women. There is a total of 6 elements. Combination of 6 taken 2 at a time is C (6, 2) = 15.

The following two restrictions apply:

1. No 2 men at a time is a valid pairing.
2. No 2 women at a time is a valid pairing.

Combinations of 3 men taken 2 a time is C (3, 2) = 3.
Combinations of 3 women taken 2 a time is C (3, 2) = 3.
The total number of exclusions is 2 x 3 = 6. If we deduct the exclusions from total number of combinations, the result is

15 – 6 = 9. The result represents the square of the number of couples: 32.

The 3 couples consist of {Wife1+Husband1}, {Wife2+Husband2}, {Wife2+Husband2}. These are the 9 pairings:

W1+H1, W1+H2, W1+H3
W2+H1, W2+H2, W2+H3
W3+H1, W3+H2, W3+H3

There are 3 cases of equal index: W1+H1, W2+H2, W3+H3 (diagonally, top left to bottom right in the matrix). They represent the *unfavorable cases*. The remaining 9 – 3 = 6 pairings represent the cases of interest: *no husband* is paired with his own wife (equivalent to *no wife* is paired with her own husband).

We can generalize for a number of *N* couples.

C (2N, 2) = {2xN x (2N – 1)} / {1 x 2) ={N x (2xN – 1)} = $2N^2 - N$

The two exclusions can be computed as
2 x C (N, 2) = 2 x {N x (N – 1) / (1x2)} = 2 x{(N^2 – N) /2} = $N^2 - N$.

Finally, deduct the exclusions from total number of combinations total husbands and wives taken two at a time:
$\{2N^2 - N\} - \{N^2 - N\} = 2N^2 - N - N^2 + N = N^2$

Total possible cases (pairings) = $\mathbf{N^2}$
Total possible homogeneous pairings (husband and wife) = **N**
Total possible heterogeneous pairings (no husband and wife): $N^2 - N = N(N - 1)$

Thus, the probability of not matching two spouses is the following rapport:

Total possible heterogeneous pairings / Total possible cases (pairings) = $\mathbf{(N^2 - N) / N^2 = (1 - 1/N)}$

That is the case of *any one* couple not being paired. But we have ***N*** couples and the original question is clear: *absolutely* ***no couple*** *is matched together*! Therefore, the final probability is $\mathbf{(1 - 1/N)^N}$.

We already saw a very similar relation in *Ion Saliu's Paradox of* N *Trials*. I noticed that $\mathbf{(1 - 1/N)^N}$ has a limit. (1 – 1/N) N = $[(N - 1)/N]^N$. I reversed and saw more clearly: limit of ***[N/(N – 1)]^N is the definition of e*** (the base of the natural logarithm). The limit of $\mathbf{(1 - 1/N)^N}$ tends to **1 / e**. There are people, indeed with training in mathematics and probability theory, who don't accept the idea that a mathematical *limit* can be reached both from the left (incrementally) and from the right (decreasingly). The limit of this probability, **1 / e,** is reached decreasingly whereas ***e*** is typically viewed as an incremental limit. The same applied to that *Ion Saliu's Paradox of* N *Trials*.

The original *Montmort's Problem* (1708) dealt with a jar containing *N* balls numbered from 1 to *N*, respectively. Let's calculate the probability that *no* ball is drawn in order indicated by its label. Montmort solved the problem in a very complicated manner, by approximating. He formulated that the probability of *no coincidences* was

$$p = 1/2! - 1/3! + 1/4! - 1/5! + \ldots$$

That formula doesn't make complete sense mathematically but does approximate the probability in question. *Warren*

Weaver (*Lady Luck*, p.139) considers that the formula *"holds accurately so long as there are at least eight men at the party, and is approximately true for smaller number of guests."*

My analysis above, using combinatorics, is a lot more precise. Furthermore, I believe that the *hypergeometric distribution probability* is a more accurate tool in this regard. We know precisely the number of all possible elements and the number of favorable elements. The hypergeometric distribution deals with four terms, and we can determine them precisely. Besides, applying the above *Montmort's formula* $[(-1)^{N-1}] / N!$, the calculations become impractical for *N* greater than 10.

On the contrary, the hypergeometric distribution probability handles easily cases such as N = 100. I was able to work up to N = 800 (*p = 1 in 2.722* or very close to **1/e**). I used my freeware *SuperFormula.EXE*, option *H (Hypergeometric distribution)*, then 1, "Standard Lotto & Keno." I worked out a few particular cases, up to N = 800. Nobody can do that by working out *Montmort's formula!*

6. Reversed Montmort's Problems or Paradoxes

Evidently, *Ion Saliu's Paradox of* N *Trials* represents the reverse of *Montmort's Problem*. The first paradox deals with at least one coincidence: *1 – 1/e*. The original Montmort's Problem deals with absolutely no coincidence: *1/e*. One probability represents the complementary probability of the other problem. There are more such problems, mostly derived from the original Montmort's Problem.

Here is what one can read in cyberforums that I have been known to frequent. Sometimes, I receive e-mails directing me to such posts.

I cannot seem to find a general solution to the problem below. It strikes me that it must be the sort of problem that has been tackled elsewhere as it can be stated so simply, but I'm not sure what it would be called and hence not sure how to search for it on Google. If you know the answer or could point me to a link I would be grateful.

Problem:

*There are **N** people, each of whom writes their name on a piece of paper and puts that piece of paper into a hat. Each person takes it in turn to remove a piece of paper from the hat, if the piece of paper they remove has their own name written on it they put it back in the hat and choose again. What is the probability that the **last** person chooses their **own** name?*

Let's try to make it as clear as possible. Let's say there are 5 persons, each with an ID number from 1 to 5 (instead of names). There are 5 tickets in the hat numbered from 1 to 5. If I was designated player no. 5, what are various probabilities that I'll draw ticket no. 5 if I am the last in line to draw (position no. 5)?

I want to know the probability that ANY player draws a ticket equivalent to his/her player ID. The probability that player no. 1 draws ticket no. 1, or player no. 2 draws ticket no. 2, player no. 3 draws ticket no. 3, player no. 4 draws ticket no. 4, player no. 5 draws ticket no. 5. We are going to dig to the deepest details.

1.1 Player no. 1 is first to draw. The probability to draw ticket no. 1 is undoubtedly $p=1/5$. No strings attached.

1.2 Player no. 2 draws after player no. 1. The probability to draw ticket no. 2 must take into account what player no. 1

did. Player no. 2 is successful if player no. 1 did not draw ticket no. 2: $p = 4/5$. Why? The first step of player no. 2's success means that player no. 1 drew any of the other 4 of 5 tickets. Then this step must be followed by the probability that player no. 2 draws one favorable ticket (no. 2) from the remaining 4 tickets in the hat. The events are simultaneous (none can be missing); therefore, the probability is the product of the probabilities of the 2 events; 4/5 x 1/4 = 1/5. The same probability as for player no. 1!

1.3 One more case. Player no. 3 draws after player no. 1 and player no. 2. The probability to draw ticket no. 3 must take into account what player no. 1 and player no. 2 just did. Player no. 3 is successful if player no. 1 did not draw ticket no. 3: p = 4/5. Why? The first step of player no. 3's success means that player no. 1 drew any of the other 4 tickets. Plus, the second step of player no. 3's success means that player no. 2 drew any of the other 3 of 4 tickets. Then this step must be followed by the probability that player no. 3 draws one favorable ticket (no. 3) from the remaining 3 tickets in the hat. The events are simultaneous (none can be missing); therefore, the probability is the product of the probabilities of the 3 events: 4/5 x 3/4 x 1/3= 1/5. The same probability as players nos. 1 or 2!

We can expand to get a simple formula of any of the ***N*** players to draw a ticket equivalent to their draw order (or index):

p = (N – 1) / N x (N – 2) / (N – 1) x (N – 3)/(N – 2) x . . . x (N – M – 1) / (N – M) x . . . x (1) / (2) x (1) / (1) = (N – 1)! / N! = 1/N

In the "player matches ticket number" case, we'll see that the probability is *1/(NxN)* for *the last player to draw last and draw the ticket with the last number.*

Let's go again with five players and player no. 5 as the pivot. We know that player no. 5 is in the last position. The other 4 players can be arranged in *4! (4 factorial) positions*. Position no. 5 never changes. The general formula in case no. 1 covers one and only one case: the ticket no. 5 is left as the last ticket for player no. 5 to draw. We have, however, 4 more equivalent situations:

2.1 player no. 1 draws ticket no. 1 in draw 1, $p = 1/N$
2.2 player no. 2 draws ticket no. 2 in draw 2, $p = 1/N$
2.3 player no. 3 draws ticket no. 3 in draw 3, $p = 1/N$
2.4 player no. 4 draws ticket no. 4 in draw 4, $p = 1/N$

Overall, we have 5 cases. The probability that *only one* of them will occur is formula 2:

$$p = (N - 1)! / N! / N = 1 / (N * N) = 1 / N^2$$

If there are 10 players, the probability that player no. 10 will draw ticket no. 10 if drawing last (position no. 10) is $p = 1/100$. I verified. Everybody can verify. The closest tools are more often than not the most useful tools. ***MDIEditor and Lotto WE*** is a big piece of software I also wrote. Click on the menu bar *Lotto, Odds+Random Lotto*. Select a game of *5 with 1 number per combination*. The combination generation is very fast. Stop it quickly. You have to count the occurrences manually using the *Goto Line* command. If number 5 is in a multiple-of-5 position, count it as a success. I did get 5 of 95, 5 of 110, and 5 of 105 successes for a *1-of-5 game*. It must be that the formula 2 is correct. At first, I believed the probability was *1/N* (not $\boldsymbol{1/[N^N]}$). Remember, the probability is calculated as *exactly no. M draws ticket no. M*. We need to count all such occurrences and then average them out.

7. *No Problem!* There Are Solutions to Every Probability Problem

Obviously, it is impossible to even list all possible problems related to probability. The paradoxes are included!

I believe that this book offers an effective path to solving *everything* life throws at us, including problems of probabilities or odds. Even if the odds are stacked high against us, at least we know.

The most effective method is to break it down to the *fundamentals*: define the elements. What is favorable, and what is unfavorable? Are there *neutral* cases such as a *push* (or a *tie*) at the blackjack table?

Next, *count* or *enumerate* the elements: the *favorable cases* and *total cases.* No doubt everything is more difficult at this step. We do have an effective tool however. We know a lot about *combinatorics* and comprehensive knowledge of the four types of *numerical sets*.

Furthermore, we also have an extraordinarily powerful tool: *computer software*. I presented here only my software. Probably I have written the largest database of software in the probability field. But I am not the only one, and I will not be the only one. My software can also be used as a model for future programs. You, the reader of this book, might be one of the next programmers. A number of my programs will be somehow released to the public in the form of a source code! (Some of my software source codes are already available from my Web site, SALIU.COM.)

I learned well this lesson: theory can be really tricky. The humans are biased. We fall in love with our own ideas and abandon them reluctantly. As for me, I no longer feel bad if I realize I was wrong a while ago. I test just about everything by running other people's software or my own software. If nothing is available, I fire up my compiler and write software that tackles the problem. I feel no pain if the software proves to me that I was wrong *a while ago*!

And thus we reach that part in this book that deals with *real applied probability theory*. It will show plenty of personal examples of how I applied my probability knowledge to very practical purposes. Namely, my casino gambling and lottery playing are entirely founded on theory of probability. Gut feeling has not been part of my play for a long, long time. The Fundamental Formula of Gambling cured me—although it hasn't made me a millionaire, not even a rich person . . . yet!

Chapter VIII

GAMES OF CHANCE: REAL-LIFE PROBABILITY THEORY

1. Games of Chance and Probability Theory: A Historical Perspective

We talked in chapter 7 of another philosopher and mathematician: *Pierre Rémond de Montmort*. Montmort's reputation was established by his book on probability *Essay d'analyse sur les jeux de hasard* (1708). The book is a collection of combinatorial problems and a systematic study of games of chance. I shall do the same here: analyze games of chance from our day's perspective.

Les jeux de chance or *games of chance* signed the birth certificate of theory *of probability*. We already know when and how it all started.

Blaise Pascal was posed with two famous problems by a noble named *Chevalier de Méré*. M. de Méré was a gambler, especially in games of rolling the dice.

There are historians who move the birth of probability theory by one century earlier (the sixteenth century). One notable

Italian lawyer, with special mathematical skills, wrote the first treatise of a new branch of mathematics related to probability. He was *Gerolamo Cardano* who wrote *Liber de ludo aleae* (*On Casting the Die*). The book was written in the 1560s but not published until 1663. It might be one reason why *Blaise Pascal* is credited as the *father of probability theory*, not *Gerolamo Cardano*. It is more fact than anecdote that Cardano solved his financial problems with the help of his extraordinary skills as a gambler (chess player as well)!

Alea is Latin for *dice*. In many cultures, *random* events, as studied by probability theory, are called *aleatoric*. Theory of probability is sometimes known as *mathematics of aleatoric phenomena.* I would suggest a new name for theory of probability: *Aleatora* (or *Aleatoria* or even *Aleatorica*).

Other luminaries and pioneers of probability theory analyzed games of chance and discovered new mathematical laws, rules, and formulas.

The two *de Méré problems* created an exchange of letters between *Blaise Pascal* and *Pierre de Fermat* in which the fundamental principles of probability theory were formulated for the first time (by most accounts). Most historians do not give credit to *Gerolamo Cardano* for his contributions to the mathematical analysis of aleatoric games. And thusly, it is widely considered that no general probability theory was developed before the famous correspondence between Pascal and Fermat.

The Dutch scientist *Christian Huygens*, a teacher of Leibniz, learned of this correspondence and shortly thereafter (in 1657) published the first book on probability (as so defined by many). Entitled *De ratiociniis in ludo aleae*, it was a treatise on problems associated with *gambling*.

As Julius Caesar put it, *"Alea jacta est."* The die has been cast.

Because of the inherent appeal of *games of chance*, *probability theory* soon became popular, and the subject developed at a rapid pace during the eighteenth century. The major contributors during this period were luminaries such as *Jakob Bernoulli* (1654–1705) and *Abraham de Moivre* (1667–1754).

In 1812, *Pierre de Laplace* (1749–1827) discovered a host of new mathematical techniques in his book *Théorie analytique des probabilités*. Before *Laplace*, probability theory was solely concerned with analyzing mathematically *games of chance* or *gambling*.

For more, see *Calculus, Volume II* by Tom M. Apostol, published by *John Wiley & Sons* (1969).

2. My Contribution to Probability Theory and Games of Chance (Gambling, Lottery, Sports)

Indeed, the *games of chance* or *gambling* have an inherent appeal. Gambling represents a very popular form of entertainment the world over. All humans have hopes and dreams. Gambling is a huge business as well. The phenomenon has also *stigma* attached to it, especially placed by overly religious persons. And thus, gambling is a *Kingdom of Sinners*, and there is *Sin City* as the capital (*Las Vegas, baby*!).

I have early memories regarding games of chance. I remember when I was a young schoolboy in my native village by the Danube River in Romania. My grandparents, parents, sister, and I enjoyed playing cards. It was the best form of entertainment available to us, the peasants. Radio was very rare, and there was no television around. The homes only had speakers connected to the village power radio (run by the Communist government).

I was pretty good at playing cards, and most people credited my special mathematical skills for my successful "gambling."

Let's say I started scholarly with sports betting when I was a student in high school. There was (still is) a form of gambling run by the government: *sports prognostication*. The prizes were very high, by local standards, when the prognostication used Italian soccer teams. The soccer (real football or European football) in Communist countries was quite predictable (and subject to corruption also). The host teams would win very, very often. Also, the elite teams (those representing the armed forces and the repressive apparatus of the police and secret police) were much more successful than the rest of the competitors. In contrast, the Italian soccer championship was very competitive. Also, the results in the Italian championship had a high degree of randomness.

I had not discovered the Fundamental Formula of Gambling (FFG) at that time. Especially, I had not discovered the fundamental concept of *degree of certainty*, *DC*. My soccer-prediction strategy was statistical. The method will be presented in more detail in the next chapter. I recorded the results of every Italian football team, separated by *home results* and *results on the road*. When I analyzed a match, I put together the wins of the home team and the losses of the visiting team; the other group combined the losses of the home team with the wins of the visiting team. The greater of the two groups, by an established percentage margin, determined the *prediction*. If the two groups were equal or very close by a percentage margin, the prediction was a tie. Sometimes, two predictions were made such as *home win* (1) or *tie* (x).

On average, my method would predict 9 or 10 matches (out of 13). That was not satisfactory however. (The yield of the method is quite profitable for American sports betting; however, a 65% success ratio can lead to substantial profits.) The payouts in Romanian sports prognostication started with at least 11 games predicted correctly. I did achieve that a few times, even 12 correct results sometimes. I also had predicted the maximum 13 very few times. Unfortunately, the payouts were not great when I had 12 or 13 (or else I might have not been here, writing this book!).

There were also lottery games (still are) during my life in Communist Romania (everything was run and owned by the government). Again, my lottery strategy was primarily statistical. I analyzed the *frequency* of the lottery numbers and the *pairings* of the numbers. Later, while in USA and an accomplished computer programmer, I further developed that early lotto strategy to what is now known as the *lottery wonder grid*. My winnings existed, but they were not really large.

One day, while at work (I was an economist), two of my colleagues pleaded with me to run my lottery strategy. The promise was that I would participate in playing without money as I was the one who did most of the work! The two guys (also economists) took my sheet with the lottery combinations most likely to win. And win they did! It was the second-tier prize, but it was a large one (by national standards). The two guys split the profit between them. They didn't give me a penny—I mean, nothing. They claimed that I hadn't paid them my share of the ticket's cost!

Life goes on. I landed in the United States in 1985 as a refugee. It was only a few months before I played an American lottery game for the first time. I didn't have a vehicle at that time—couldn't afford one. Good Samaritans gave me rides when I needed to buy food or other major necessities. One day, one of my fellow farmworkers gave me lottery materials (a list of past drawings, tickets for various games, etc.). He invited me to play with his group. The Puerto Rican later told me that he had figured out I was a very smart guy; therefore, I might be able to figure out how to win the lottery! I am not kidding! We won the very first time I played (the third prize). Each one of them played my lotto combinations as well. Everybody was happy. We played again next drawing—and we won again! The *lottery wonder grid* looked miraculous to some. The fellows were lowly educated, but they understood what my lottery strategy was about. That is, recently drawn numbers tend to repeat more often, including recent pairings of numbers. It was related to the *FFG median* as I would spell out several years later.

As poor as I was, I bought a tiny personal computer as soon as I had saved enough money (from working as a farm laborer). The *electronic toy* was a *Timex Sinclair*! My first—and most important—project was *generating lottery numbers*! The tiny instrument didn't allow me to do much of any statistical analysis of past lotto drawings. Still, I was able to do something I consider an important discovery. I asked around for months and months with regard to source code to generate lotto combinations in lexicographical order. It seemed as if nobody had a clue. I stopped asking and started to work my brains off.

By the end of 1985, I had the lotto-combination-generating algorithm working perfectly! I stepped up and bought a new computer. It was a bargain all right (a discontinued *Atari 800*

XL). It was a powerhouse by my standards! It had 64 KB of RAM (compared to the previous 4 KB of Sinclair). I was able to develop my lottery algorithms to new heights. My algorithms were now real lottery programs. I could also do statistical analyses of a few recent lotto drawings. It was not easy though. My *Atari 800 XL* did not have a disk drive, and I couldn't find one (the home computer was a discontinued model). It very much looks nightmarish now, but I had to type the source code in BASIC every time I wanted to run a program! Believe it or not, I enjoyed the process. One must enjoy the process if one wants to perfect the process and thusly perfect himself/herself.

I played a few more times with the same Puerto Rican fellow and another one (who was also a great singer). We did win a few times, but only the third-tier prize. We felt good, and our hopes grew higher. We fantasized about winning the *big one*. The older guy, the Puerto Rican singer, scared me one day. He said that some people would be unhappy with my mathematical lottery strategy and computer software. As I was riding a bicycle to go to work, some vehicle might kill me on purpose and make it look like a traffic accident! In the fall, I rode early in the morning when it was still dark. Of course, I'm still here as I write this memorable book.

My *Atari 800 XL* was working at home while I was working hard pruning trees in the orchard. My program checked a dozen or so past lotto drawings—the most recent ones. My new strategy was based on eliminating various groups of numbers from the previous drawings. For example, two groups of numbers from the most recent two to five draws and all groups of three numbers from all past drawings. I would have liked to use thousands of past drawings and eliminate many more groups of numbers. What a difference a decade makes! My lotto software works now with hundreds of thousands of lotto drawings—even millions (real draws and simulated drawings or randomly generated combinations).

I came back home from work. My first task was to stop my computer and write down the last lotto combinations generated. I needed some time to fill out the lotto tickets—we played thirty-six lotto tickets. The first Puerto Rican fellow, the one who would drive me to the food store and the lottery outlet, was in a hurry that day. He was separated from his children. One of his sons was visiting that day. The caring father decided that we had to play the lottery tickets from the previous week instead of waiting for me to fill out the new tickets (and possibly make mistakes). It was Friday, the day I would buy groceries too. I wouldn't buy groceries that day (it was going to happen a couple of days later).

One of the lotto combinations my computer generated that Friday, February 13, 1986, hit the jackpot. The jackpot was in excess of three million dollars—and nobody won it! I had the winning combination only on paper, in my apartment. The ticket was not filled out, and we had played the tickets from the previous week. I lived a tragedy, much more aggravated by the mocking workmates for many days to come. The story is well documented at my Web site. You will find a few pages that deal with that incident.

I wrote lottery software at a scale far larger than before. I stepped up to IBM-compatible PCs and compiled my programs with more and more efficient tools. I had less and less time to dedicate to actual lottery playing. Meanwhile, the Fundamental Formula of Gambling (FFG) grew more and more clearly inside my brain. When I figured out the all-important concept of *degree of certainty*, my gambling theory reached new heights. I expanded the concept to *horse racing* and the huge phenomenon named *casino gambling*.

I entered an *offtrack wagering* for horse racing early in 1995 for the first time. My first encounter was a successful one, as it happens more often than not in my case. My first horse racing betting method will be presented in more detail in a future chapter (after sports betting).

In the Christmas season of 1995, about which I do not care anymore, I visited Atlantic City for the first time. It was my first casino experience. Again, it was a successful one. I had gathered undeniable evidence that the Fundamental Formula of Gambling is the most efficient tool in winning at all *games of chance*. My predecessors in the field of probability theory mostly *analyzed* games of chance or gambling. Me, I have analyzed games of chance and have discovered new laws of my own. At the same time, I consistently applied my theory in real life—play, that is.

How certain people react toward you says a lot about your ideas. My first Internet experience came in 1997. My Internet service also provided free space for a small Web site. An introduction to Socrates was the first thing I published on my new Web site. The other thing I published was the Fundamental Formula of Gambling (FFG), the table derived from it, and the interpretation of FFG. Many Net surfers reacted positively to the formula and its interpretation. A group of other people reacted with intense hostility. Who were the haters and the attackers? It didn't take long to figure them out. Most of them were related to casinos and the gambling industry.

There were a few individuals with interests similar to mine: probability theory and gambling. Their hostile reaction attempted to intimidate me, to make me fold up shop and run away. Then they would republish my work in their own words and lay claim on the discovery of the Fundamental Formula of Gambling. One alleged educator, after a few unsuccessful tries at intimidating me, resorted to a weird form of begging.

He begged me in the name of his students that I remove FFG from cyberspace!

By and large, however, the hostile reaction, sometimes intensely so, toward FFG has always come from the casino high ranks. There is an incident (2001) that still attracts a lot of attention at my Web site. The chairman of the largest casino company, MGM Grand, took my message board by storm. You can read the whole story and more.

> *I am currently the Chairman of the MGM Grand, and I was advised of this site by way of a memo sent to me. I have dealt with EVERY aspect of the casino floor at one time or another and I can attest that Roulette is a completely random probability.*
>
> *There are a few obvious functional problems that can occur (such as wheel bearings, etc.)—and those problems are offset on a daily basis. There is also monitoring system that will alert us when an extenuating circumstance of odds exists within the game.*
>
> *There is nothing mathematical about it. Don't waste your money purchasing supposed "get rich" programs based on Roulette. They simply will not work. Some of these programs are costly and also impractical, in the sense that they require you to enter a casino and keep track of numerical data.*
>
> *Although not illegal, we have (and reserve) the right to refuse service to anyone. If a spotter or floorman notices you writing things down, you will most likely be asked to play another game or asked to leave the casino floor. Failure to comply will result in being arrested for trespassing.*

This is simply the way things work. I'm not trying to come off as being harsh, but I want to make sure you are fully aware of the situation.

Evidently, mathematics, probability theory, and FFG do scare some entities, casinos especially. It inspired me to coin a catchy phrase that many people really love, as they communicated to me:

[FFG is the reason why] God fears mathematics while Einstein hates gambling.

(Remember, Einstein stated that *God doesn't play dice with the Universe*. Or *God doesn't care about our mathematics.* In truth, the Universe is a dicey game to an extreme! Einstein did not grasp the concept of probability at all. For him, *Everything* was *Perfect Order* created by *God*. *God* received a new name: *Light*. I personally consider Einstein to be one of the most intelligent mystics of all time. His fame is largely the result of the Western establishment comfort in a brilliant mind "proving" the existence of God. It resembles the agnosticism of medieval Europe. Only God knows everything because God is absolute and perfect. Humans cannot achieve a high degree of knowledge because God's ways are unknown to man. Mathematics, a human illusion, is impotent in acquiring knowledge and truth! Some luminaries, such as Galileo Galilei, accepted the humiliation of the compromise in order to survive. Another group of luminaries, such as Giordano Bruno, did not accept the compromise at all and ended up burning at stake.)

In 2003, an extreme case happened to me while gambling in Atlantic City. Not only would I not lose money, in fact I would also win money in any casino I entered. And not one but two games: blackjack and roulette. An angry floor manager stopped me from playing. The blackjack dealer was ordered to stop the game altogether. I filed a complaint with the state government watchdog. The state agency sided with the casino, of course. I did not have the financial resources to hire a lawyer and file a lawsuit in federal court. Again, that incident is presented in great detail at my Web site (including the names and badge numbers of casino personnel).

Even as we speak, the casinos find out immediately who I am as soon as I open a notebook and write down the previous events (blackjack hands as Win [W] or Loss [L], roulette spins, and the results as W/L). Also, a student in gambling management at a university in Nevada published the fact that I am close to the top of a casino blacklist. The guy at the top of the blacklist is an actual cheater, using video devices to "see" the cards in a deck.

Certainly, the casinos have a point. On one hand, they denigrate persons like me, labeling us insane. Simply talking about mathematics and its application to gambling attracts the label of insanity. Of course, I don't care a bit about that. Au contraire, it strengthens my conviction in the validity of my theories. If it's insane, why do they care?

On the other hand, if all gamblers would be like me, following probability theory instead of guts, then there would be no casinos! No casinos at all! The casinos exist only because the patrons (the gamblers) lose. If every player wins, gambling would be a huge charity. But who's going to fund it? Therefore, the casinos are almost compelled to react with intense hostility to *smart gambling*. That is, as in the very beginning of probability theory, gambling based on mathematical analysis.

I want to write here right now a *disclaimer*. I make this disclaimer not out of fear of the law but because of my consciousness. I do not make you, the reader of this book, any promise. I do not promise you that you will be a winner automatically after you read and even study thoroughly this book and my strategies. Nor do I make a commitment to assist you in using my software. My software is free to use unconditionally (after paying a nominal membership fee to download). There is also plenty of documentation—both companion to the software or Web pages—at my site.

I present here various forms of mathematical analysis of a comprehensive collection of games of chance. The mathematical foundation of the strategies and systems presented is undeniable. Not even the fiercest of my foes and detractors have been able to prove me wrong mathematically. It is up to you—how you understand, comprehend, interpret, and apply my theories, strategies, systems, and software. But I can guarantee with a high degree of certainty, DC, that many readers will definitely benefit from reading attentively this book. They do gain valuable theoretical knowledge in the first place. They also acquire valuable tools in becoming significantly more successful in handling tasks that seem to be *mission impossible* (e.g., winning consistently at lottery or gambling).

And as a repeat reminder, this book should be always intertwined with my Web site. There are plenty of pages with anecdotal facts that have no place in this book. It would turn too voluminous. But facts are facts, and they can be verified. As Ronald Reagan and Mikhail Gorbachev put it in the 1980s, *Trust, but verify!*

Perhaps a future edition of this book will incorporate, on CD or DVD, my entire Web site and my entire collection of software.

Chapter IX

THEORY AND STRATEGY OF SPORTS PROGNOSTICATION OR BETTING

1. The Statistical Approach: American Football Betting and Italian Soccer Pools (1, X, 2)

We know from the beginning of this book that the *past* does count in all random events (in everything, that is). For one, the past is represented by the *number of trials*—one of the three essential elements of the Fundamental Formula of Gambling (FFG). This is how I started my *gambling theory*. Statistical analyses can discover *tendencies*. A tendency is, simply put, a repetition of something across a number of trials.

Back in Romania, I loved to play soccer pools: 1, X, 2. The numeral *1* represents a win for the home team; *x* is a tie (very rare in the American football, very common in the Italian calcio leagues); *2* means a win for the visiting team. I needed to devise a strategy to give me the best results based on statistics. The system was best suited for the Italian soccer championships (*Serie A* and *Serie B*). The Italian soccer games have a high level of unpredictability. In most other national championships, the results are more easily predictable (the home teams win in most cases).

I did not write stand-alone software for this type of gambling mathematics. Instead, I created two Excel 95 spreadsheets. They are still available from my Web site although data is outdated now. The user can update all figures however. Long live the Internet!

The European football (soccer) spreadsheet is named *Toto1x2.xls*. It is based on real data from the 2001 Italian football championship *Serie A*.

The American professional football spreadsheet is named *NFL2001.xls*. It is based on real data from the 2001 NFL regular season.

I will present now an abbreviated form of my system applied to American football (specifically the NFL). As I do not offer a computer program at this time, you can apply this strategy manually (pencil and paper) or edit the spreadsheet.

You will keep records for each team, separated in two files (spreadsheets): record at *home* and record *on the road*. Using the overall records has much lower relevance in American professional football. The betting is based on a parameter named *point spread*. A team is expected to win all right. But it must win by a specific margin of victory (the point spread or simply the spread).

You will work with these two very important elements: *Home Team at Home* and *Visiting Team on the Road*.

In this abbreviated version, you will use only the results (win/loss) and the points scored/allowed. Add the wins for the home team at-home with the losses for the visiting team on the road. Do the percentage. Normally, a percentage over 75% is a good indicator for a home-team win. The average scores can offer additional useful info. In the case bellow, the game was very close to call: just a slight advantage for the Home Team. But the Visiting Team had a realistically good chance to win (based on points *scored* and points *allowed*).

Using my strategy for the Italian soccer leagues, I predicted 9—even 10—(out of 13) games (70% or 77%). That is, I predicted with only one prognostication sign: 1 or X or 2. Then I would use two signs for close games (1X or 12 or X2, even 1X2). The soccer prediction is based on the *win/loss* parameters. The scores are disregarded. Thus, the soccer betting appears to be simplified compared to American football betting.

I lived as a refugee in Yugoslavia for eight months, between 1984 and 1985. That's when I had the most success playing soccer pools. I helped also the Yugoslav office of the United Nations High Commissioner for Refugees (UNHCR). I translated statements and documents from Romanian into English. I enjoyed doing the job. I helped others, and I helped myself by practicing English and getting to know more about the *Homo Computing_beast* experience. It is the name I find to be most truthful for this extraordinary species.

UNHCR promised to pay me. I would hear *"Sutra! Tomorrow* [in Serbian] . . . *Next week* . . . " The main problem I had then: my shoes were real bad. Granted, they helped me to cross the border. But I had reached a point I needed a pair of better shoes. I turned to an old friend, gambling. *Specifically, the soccer pools*. I won a couple of times playing the Yugoslav soccer championship. As in most countries, there were not many surprises to lead to big prizes. Then a beautiful medieval winter set in. Every country in Europe would use only the Italian soccer games for their pools. I hit the second prize. Actually, I predicted all the games, but I applied a *minus-1 system* (a wheel). Anyway, the prize paid good money, especially by the standards of a shoe-challenged refugee.

I didn't have time to do anything after cashing in the prize. Just about everybody in my "politically correct" entourage wanted a big party. The refugees live under tremendous stress in case you didn't know that. So I said, "Yeah, let's do it, we need it!" I thought everybody got drunk that night, but not everybody did because somebody stole all my money overnight! Nothing was left in my pockets after the big party! Truth is I arrived in America wearing bad shoes, perhaps to prove to myself the American dream. *"Give me your poor, your shoeless, etc."*

Let's change gears and take a look at a real-life game in the NFL and the stats for the two teams.

Home Team: At-Home

Week	***Wins***	***Losses***	***Points Scored***	***Points Allowed***
Week 1	1	0	20	13
Week 2	1	0	13	10
Week 3	0	1	20	27
Week 4	1	0	21	14
Total	***3***	***1***	***74***	***64***
Average	***75%***	***25%***	***18.5***	***16***

Visiting Team: On-the-Road

Week	***Wins***	***Losses***	***Points Scored***	***Points Allowed***
Week 1	0	1	20	31
Week 2	1	0	21	14
Week 3	0	1	16	35
Week 4	1	0	24	13
Total	***2***	***2***	***81***	***93***
Average	***50%***	***50%***	***20***	***23***

Predicted Result Based on Wins/Losses:

Win for Home Team: 3 (wins at home) + 2 (Visitor losses) = 5 of 8 (62%)

Win for Visiting Team: 1 (Home losses) + 2 (Road wins) = 3 of 8 (38%)

Home Team
Average points scored: 18
Average points allowed: 16

Visiting Team
Average points scored: 20
Average points allowed: 23

Predicted Score: 20–18 in favor of Home Team

The prediction for this NFL game was derived from four games only. The success ratio increases with the increase in the number of games. The results get better in the second half of the regular season (after week no. 8). Using data from previous seasons is very often misleading. The teams in the American sports change constantly as a result of the *free agency*. Players move from one team to another. Some players have a real impact on the game.

The success (correctness) rate of *SPORTS.XLS*, the Excel spreadsheet for sports betting, is around *2 out of 3*. Actually, it is higher at picking *straight winners* or *pick-'em*. It can still hit at least 60% when applied to *point-spread* betting as in the American football.

A 60% success rate translates equivalently that a player can pick one game at a time and be successful in 60% of the cases.

If picking *2 games*, the probability goes down to .6 x .6 = .36 or 36%.

If picking *3 games*, it's .6 ^ 3 = 21.6%.

A *five-team parlay* can expect a success rate of .6 ^ 5 = 7.7%.

The most favorable case is for *straight-up* betting: one game only. The house advantage (edge) is also the most favorable: *10%*. That is, a 1-unit bet returns a .9-unit payout. Usually, American football bettors pay $110 for a $100 win.

A five-team parlay offers, at best, a 25% house advantage. Ten or more game parlays go to over 90% in house edge! It is outrageous!

If betting straight up 10 times (or weeks, as in NFL pro football), the player should expect 6 wins and 4 losses. The four losses result in 4 units. The 6 wins amount to +(6 x .9) = +5.4 units. The net win is 1.4 units. If betting $100 at a time, the net profit should be $140 over a period of 10 weeks. Just for fun unless thousands are bet per game.

A probability of over 50% yields also longer winning streaks and shorter losing streaks. The mathematical player can draw a big advantage from that probability feature. The nonmathematical bettor should keep in mind that an over-50% probability tends to repeat in the very next trial; a bet increase is favorable. Conversely, a 60% probability event will not lose very often more than three consecutive trials. Again a bet increase after three consecutive losses is good money management.

My spreadsheet strategy applied to American football prognostication was pirated even at a Microsoft site! One visitor to my sports forum advised me of the incident. Go to *http://lotterygambling.phpbbnow.com/viewtopic.php?t=111.*

Here is an excerpt from the Microsoft site (the URL is *http://office.microsoft.com/en-us/excel/HA011246011033.aspx?pid=CL100570*):

> *Starting in row 36, columns C and D contain the team code number (listed in B2:B33) for the home and away team for each game. For example, the first game (listed in row 36) is the San Francisco 49ers (team 28) playing at the New York Giants (team 21). Column E contains the home team's score, and column F contains the visiting team's score. As you can see, the 49ers beat the Giants 16-13. I can now compute the outcome of each game (the number of points by which the home team beats the visiting team) by entering the formula =E36-F36 in cell G36.*

That page was published several years after I released my sports betting theory and spreadsheets (2000).

2. The Randomness Approach: Resemblance to Lottery Games

How about *random* play like *quick picks* in lottery? It could help the biased sports-betting player. Many sport bettors have a bias toward favorite teams, including football. Reality is that only a tiny minority of teams are dominant. Therefore, most biased players are losers. They would be better off choosing the winners randomly. I offer free software to pick sports results in a random manner like the lottery. This section lists the most important free programs that generate random sports picks—as *numbers* and also *team names*.

A program I wrote in August 2003 has become a big success: *PermuteCombine.EXE*. The freeware generates permutations, arrangements (some say permutations of *N*, *M*, like exactas and trifectas at horse racing), combinations (like in lotto), and exponents (like in the pick-3 lottery and soccer pools 1, X, 2).

The program generates numbers or *words* (such as names of sports teams). There is a sample text file—TEAMS. NFL—with all 32 NFL football teams. The option applicable to the American football is "Arrangements, Words, Random." The option applicable to soccer pools (1, X, 2) is "Exponents, Words, Random" (use the sample file 1X2.TXT).

It's best to run a random option several times in a row. Common sense tells you that hitting in a first try is a very rare event in gambling. The high-eyebrows translation (probability theory) is that the degree of certainty, *DC*, rises with the increase in the number of trials, *N*, while the probability, *p*, is always constant. Repeating the combination-generating process affords to you compression of time and costs.

Here is an interesting fact I discovered. Please relate it to the role of *standard deviation* in random events (the watchdog of randomness, chapter 4). I applied this NFL strategy live at my Web site. That is, I published the predictions days *before* the games were played. I recorded the results after the games ended. The new page was published with the new data. I had *misses*, and I had *wins*. The project covered almost three regular seasons in the National Football League (NFL). The results were quite impressive: 27 wins out of 41 weeks. Success ratio: 65.9% (2 out of 3 times). See the page at SALIU.COM, random-picks.html.

I applied the strategy to five games. There are two outcomes in rapport to the betting line. Either choose the *favorite* to cover the spread (win by the point spread) or bet against it (take the *underdog* to defend the spread). Total amount of possible combinations is therefore $2^5 = 32$.

The Fundamental Formula of Gambling demonstrates that, in a majority of cases, the winning combinations will come from the *FFG median zone*. The FFG median zone and the normal (Gauss) bell share a large percentage of elements, but the areas are not identical. Percentage of 2^5 combinations within the FFG median bell is 38% (or 12 lines of 5 teams each from index no. 11 to index no. 22).

My site also publishes the schedule of the National Football League (I have pages starting with the 2004 season). When this project was running, I copied the schedule from the official site of the NFL. Let's say that a particular week had 16 games scheduled. I ran *PermuteCombine.exe*, the *Combinations-Numbers-Random* option for 16 numbers, 5 numbers per combination. Total combinations to generate was 22 (the median calculated by the Fundamental Formula of Gambling for a probability equal to 1/32). The combination no. 22 for that particular run was 4, 5, 7, 8, 11. In the case described at my site, you can see what games those 5 numbers represented in that week's schedule. The schedule starts with game no. 1 (of course) and ends with game no. 16. For example, game no. 4 that week was *Texans at Jaguars*.

The winning combination for that particular week was Jaguars, Colts, Chargers, Vikings, 49ers, or lexicographical index 20 (a winner—inside the FFG median area *11-22*).

The accounting or the auditing now
Total games to play every week: 12
Total weeks played: 41
Total games played: 41 * 12 = 492
Total winning weeks: 27

Payout for a winning five-team parlay: 25
Total winnings: 27 * 25 = 675
Net profit: 675 – 492 = 183

Not long after my project, most bookies lowered the five-team parlay payout to 20-to-1. The total winning would go down to 27 * 20 = 540. Still a profit (48 betting units). Since the winning percentage is well above 50%, increasing the bet after a loss offers a great benefit overall. I would even start martingaling after 3 losses in a row! There was one occurrence of 3 consecutive losses during the 2002 season. The martingale betting could have started after just 2 consecutive losses, and the profit margin would have been phenomenal!

The sport betting area is the least covered by my software. Nonetheless, I still provide the largest collection of sports prognosticating software in the world. I list here the most significant programs that you can easily obtain.

• *Sports.xls.* Sports betting software, real data from the 2001 NFL season (Excel 95 spreadsheet).

• *NFL2001.xls.* Sports betting software, real data from the 2001 NFL season (Excel 95 spreadsheet).

• *Toto1x2.xls.* Sports betting software, real data from the 2001 Italian football championship *Serie A* (Excel 95 spreadsheet).

• *SkipSystem.exe.* Gambling and lottery software to automatically create lotto, lottery, gambling, sport-bet systems derived from skips and based on the FFG median.

• *FrequencyRank.EXE.* Software to generate frequency reports, including wins of the NFL teams (against the point spread).

• *BetUS32.EXE*. Intelligent random-combination generator for sports betting, the American way.

• *AmBet.EXE.* Much like BetUS32.EXE but with a twist! You can add *W* or *L* at the end of a team name based on the results of the previous week.

• *BellCurveGenerator.exe*. It generates combinations within the FFG median bell, including for NFL games and soccer pools (1x2).

The download location for sport-bet software category is *saliu.com/free-sports-betting.html.*

Chapter X

THEORY AND STRATEGY OF HORSE RACING WAGERING OR BETTING

---

1. The Paper-and-Pencil Approach: Horse Racing at Offtrack Wagering (OTW)

In 1995, I entered for the first time an offtrack-wagering (OTW) parlor. The facilities show horse races from across the world, although horse tracks from North America are covered most extensively. The OTW had been in the news for some time. That OTW facility in York, Pennsylvania, was opposed by a number of people, mostly religious activists, elderly, and self-proclaimed crime stoppers.

Again, the stigma placed on gambling! People were also afraid that the organized crime would invade the territory, and thus the crime would turn into a real plague!

Those puritans claimed that offtrack wagering (OTW) would bring addiction and high crime to their neighborhoods. The reality today is a far cry from such a fear. The OTW facilities are nice and clean. They also have good restaurants where friends and family meet for dinner. The OTW companies actually brought the crime rate to zero in their vicinity. The places were very crowded in the beginning. Now they are half full more often than not. The current patrons would have otherwise spent (and lost) much more money playing the government-sponsored lotteries (with monstrous odds and house edge).

There is no more illegal betting on horse racing. Since it is all done in daylight, the government actually collects some extra money in taxes. Why are other forms of gambling banned? I can see no other reason but to encourage illegal gambling! And anything that's illegal increases the crime rate and addiction, to say nothing of lost billions in uncollected taxes. *Big Brother* should regulate instead of ban.

I remember how I applied for the first time my *pick-3 lottery method* to *horse racing* that day in March of 1995. I remember it was the Sportsman's horse track (Illinois, USA) that I focused on. The name was appealing. I had tracked two previous races at Sportsman's. I wrote the results as horse numbers only on the margin of the OTW program. I was watching also the track on the TV set at the left and the races at the track covered by the TV on the right side of the Sportsman's TV. I was sitting comfortably at the bar, sipping a beer.

So I had 6 winning horse racing trifectas on a piece of paper. I didn't want to play any of the previous numbers in the same positions. I only was careful not to play really long shots in position 1 (to win). I came up with a few combinations. One of them was the straight winning *trifecta*: 6-1-3. I also played no. 6 *to-win*. Good money, but it could have been a whole lot more! It happens now and then.

I remember also the reaction of disappointment from two players next to me at the bar. One of them expressed his newly found philosophy to his partner: *"The horses have no minds, man! They don't read the program, they don't care about the odds! The horses are just numbers when you bet on them!"*

That method was really taken from my lottery software, namely the pick-3 lottery. A loyal user of my lottery software was also a horse-racing and dog-racing enthusiast. He asked me to write specialized software for horses in the manner of the lottery game that draws three digits. It was quite easy for me to transport my pick-3 lottery software to horse racing *trifectas*. A trifecta represents the first three finishers in a horse race in exact order. For example, if horse no. 10 wins, horse no. 2 places, and horse no. 7 shows, the trifecta is 10-2-7. The Canadians call the three-horse winning combination *triactor*.

How time flies! This year, 2010, I felt tired and in need to relax. Writing software and/or probability matters is not that easy, believe me! So I went back to an offtrack horse racing parlor. It was a different location than the incipient one (1995). Just for fun, not to strike it big, I applied again my horse racing betting system I first created in 1995. (I bet you I remember well that year!)

I precisely record the results, as they occur, at three horse tracks (determined by the TV monitors in the facility). I eliminate horse numbers in the same position from the recent results of the three tracks I track. Actually, last time, I could only track two tracks. It was Wednesday, and things are much slower these days.

Race no. 9 at Belmont Park. The result at the previous race was 7-2-3. No. 7 was a favorite in this race, no. 9. I came up with 2 boxed trifectas: 1-4-9 and 5-7-9. The result was 4-7-1. Had I wheeled the 5 unique numbers (1, 4, 5, 7, 9),

I would have played 10 boxed trifectas. Trifecta 4-7-1 paid over $1,000.

I still used that data plus data for the only available track remaining (Indiana). I had to wait a couple of races for more results. When waiting, it is a good idea to play the three longest shots in a race. When you hit, you get some money back. You can add to the bankroll. Anyway, the first race at Indiana (that I didn't play) had the trifecta paying over $3,000.

I attracted some attention because I only wrote down numbers. I told someone that I didn't care about the horse name or the jockey. For race no. 4, I came up with 3 boxed trifectas: 2-6-9, 7-8-9, and 6-7-11. The latter hit—after an objection delay! It wasn't big—a few hundred bucks. Trifecta 2-6-9 would have been huge—it occurred for a short time in the middle of the race! I left for home immediately. People saw my cash, and I was drinking beer.

I released a powerful upgrade to the horse racing software last July (2009). I felt nostalgic that day. My nostalgia took me back to the place where my horse racing theory started. The first result available this time was at Fort Erie. Race no. 6: 6-1-3! Groundhog day. Like in 1995, it happened minutes before I entered. Fortunately, the trifecta paid little money (fifty bucks or so).

I was not in gear to gamble. Just for a couple of cold beers to quench my nostalgia. I wrote down the results of a few races running closely to one another in time.

6-1-3
11-1-6
9-2-8
1-8-3
3-1-6

I decided that 4-5-7 were to be played to-win. None of the numbers had appeared in the first position—or other positions. It was $6 per race, 3 races in total. The first 2 had 2 of the 3 numbers in the trifectas. The final results were 4-1-7 and 4-5-9 (after a dispute). The 3rd had 5 as the winner! Not very big *to-win* payouts: $17, 6, 22. Deduct $18 and buy a cold carryout! Some races go sometimes to 50–1 winning long shots! One must be patient if one affords time and money (bankroll).

One problem at the horse racing facilities (I go to offtrack wagering OTW)—I attract attention. Most bettors around see me using numbers only. They ask me horse names, and I don't know any because I don't care about horse names or jockeys or the morning lines or the current odds for the favorite. I take the lottery approach. The horses are numbers like in lottery. The path of randomness is the best way to bet. It is mathematics. Meanwhile, other people get angry. I rarely saw a winner at any OTW. Whenever I go to horse racing, I only see and hear bettors cursing, even crying! They do rely too heavily on so-called favorites. I wouldn't like to mess with losers too often (especially if I carried a laptop)!

I enjoyed my evening at home. The weather was still hot. I sipped the beer I had won at OTW while watching a great film on my computer: *Nostalgia* by Andrei Tarkovsky.

This is the simplest and easiest-to-apply strategy for horse racing. It can be done by the *pencil-and-paper* method. Very few powerful lottery-gambling strategies can be devised as the *pencil-and-paper* method. It is not guaranteed that this method works every day. The bettor could miss a couple of days in a row. Equivalently, the method can strike multiple times during the same day.

2. The Lotterylike Strategy Applied to Horse Racing Wagering

This horse racing method is better suited for professional gamblers. They have a good bankroll, and they dedicate all their time to horse racing. You get the results and analyze them with the computer in the morning. In the afternoon and evening, you go to a track or OTW horse racing facility and play diligently and patiently across all available horse tracks.

So I recorded the results of all races from the previous day. I would go to equibase.com. I had no choice but to record all the results by horse track. It is mission impossible to write the results in sequential order.

I record data for several days. I get thousands of real results. I also have a simulated data file (randomly generated trifectas). I only look for races with at least 9 horses. The payouts are far better. I care only about trifectas that have not come out in a long time. All trifectas will come out in a reasonable time. But I only care to play trifectas that pay big. So *long shots* have to be part of the trifectas I play.

This type of betting does not take into consideration the latest results in any particular horse track. I play my trifectas in any horse race of the day in United States and Canada. Sometimes it takes several days until this strategy hits a huge trifecta. In other situations, this strategy hits multiple high-paying trifectas in one day.

The trifectas used to be much bigger before I wrote about this strategy in newsgroups. This horse racing betting method requires computer analysis and a sizable bankroll (to withstand losing days). The payouts can be big however. I heard a gentleman on the radio, after I made this strategy public, who proved he was making an average profit of $100,000 a week. The profits are lower now because a number of horse racing bettors play that way.

This high roller method can only be applicable by running my horse racing software. There is no way around it. The *pencil-and-paper* system is useless. Please be sure to read the chapters dedicated to lottery and my lottery software. My horse racing software very much works like my lottery software. The key concept is *filtering*. A *filter* is simply a *restriction*. Certain groups of numbers or *patterns* are *eliminated* or *forcibly generated* based on the Fundamental Formula of Gambling and *statistical criteria*.

I list here the most significant horse racing software titles that you can easily obtain from my Web site.

- *LotWonH.EXE* is an integrated horse racing software (consists of several programs). This package represents my oldest attempt at horse racing software. You should not run it—you should only read the tutorial. It is a good presentation of the mathematical foundation of horse racing software.

- *PickH32.EXE* is another integrated horse racing software, the 32-bit upgrade to *LotWonH.EXE*.

- *BrightH3.EXE* is, indeed, a high-powered integrated horse racing trifecta software. This software package is far more powerful, even than *PickH32.EXE*. This is the most powerful trifectas software package.

- *SoftwareHorses3.EXE* is a special horse racing utility software for trifectas (triactors) and exactas (exactors). This is the latest in horse racing software and strategy.

• *SkipSystem.exe* is a software to automatically create horse racing, lotto, lottery, gambling systems derived from skips and based on the FFG median.

• *FrequencyRank.EXE* is a software to generate frequency reports, including *straight* and *boxed* trifectas.

The download location for horse racing software category is *saliu.com/free-horse-racing.html*

Chapter XI

THEORY OF PROBABILITY APPLIED TO BLACKJACK (TWENTY-ONE)

1. The Fundamental Probability Issue: The True Odds at Blackjack

Let me start by saying that the game of blackjack has caused me the most serious problems with casinos and gambling developers, authors, and system vendors. Blackjack or twenty-one (as seen the movie *21*) is the most popular casino game and the most researched one. There are plenty of books dedicated to the so-called *mathematics of blackjack.*

There is worthiness in a few of such books or e-books. For the most part, however, there isn't much mathematics in all those blackjack studies. The heart of the matter is a worthless concept known as *card counting.*

I do have a strong interest in blackjack. It is well documented at my Web site. As a matter of fact, I consider myself the best blackjack player ever. As the great *Muhammad Ali* put, *"It ain't bragging if you back it!"* So I put the money where my mouth is. I issued a casino gambling challenge, especially at the blackjack tables. So far, nobody has dared to honor my challenge. The real casino challenge is open to any gambler, gambling author, or gambling system developer—card counting or not.

I wrote a book about the *true* mathematics of blackjack insofar as precise *probability calculations* are concerned. You might be shocked to hear, but the mathematical truth is that your knowledge of blackjack probabilities or odds is dead wrong. Everything you had known was based on guesswork, albeit educated guesswork.

To this date, the blackjack odds are the same as John Scarne calculated them in the 1950s. The computers were not the commodity they are today. And John Scarne was not a computer programmer! A far cry from that! The way he calculated the odds made sense for the first two and three blackjack cards in a round. I quote from his *Scarne's New Complete Guide to Gambling* (p.363):

> *We find that the dealer's first two cards can produce the counts from 2 to 21 in 1,326 ways.*

Indeed, combinations C (52, 2) = 1,326 two-card blackjack hands (combinations of 52 cards taken 2 at a time). That is the only thing halfway mathematically correct! The truly correct method applies the mathematics of combinatorics all right. But instead of the numerical sets known as *combinations*, we must apply the mathematics of *arrangements*. The combinations represent boxed arrangements. In this case, C (52, 2) = (52 * 51) / 2 = 1,326. Arrangements A (52, 2) = (52 * 51) = 2,652 (or double the amount of combinations). Hence, we played cute and said half true for the blackjack combinations case!

> *We'll discover that we need to know, however, and avoid most of the fractions, if we multiply 1,326 x 169 to get a common multiple of 224,094.*

Now that's a big mystery! How did Scarne come up with that *169* factor? Well, that's what they call an educated guess or *guestimation*! John Scarne, *requiescat in pace*, didn't have a clue, mathematically speaking. He has never explained how he came up with that 169, kind of a new number of the beast!

In order to calculate the probability *precisely*, we must generate all the elements (blackjack hands) in *lexicographical order*. Nobody even knows how many hands are possible, as their size varies widely: from 2 cards to 10 cards (for 1 deck)! When 2 or more decks are employed, the blackjack hands can go from 2 cards to 11 cards.

Of course, there is a lot of blackjack software out there! But all that software belongs to the *simulator* category! That is, the blackjack hands are dealt *randomly.* Based on the well-known-by-now *Ion Saliu's Paradox*, random generation does not generate all possible combinations as some elements repeat. So we can never calculate the probability precisely based on random generation. If there are *334,490,044* total possible complete hands in blackjack, only 63% will be unique, and 37% will be repeats—*if* we randomly generate 334,490,044 hands.

I rolled up my sleeves again. I had started years ago a blackjack project to generate all possible hands. It was very difficult. I put it aside and forgot about it as other projects felt more compelling to me. I found the project this year (2009) and also the code to generate sets from a list. In this case, the list is a 52-line text file with the values of the blackjack cards from the four 2s to the sixteen 10s to the four aces. That's a stringent mathematical requirement. The deck of cards must be also ordered lexicographically if we want to correctly generate all qualified sets in lexicographical order.

Let's make this analogy to lotto combinations. Generating lotto combinations in lexicographic order is far easier than generating blackjack hands. The lotto combinations have a *fixed length.* For example, a lotto 6/49 game consists of 6 numbers per combination from 1 to 49. There are 13,983,816 total possible elements for this type of lotto game.

Incidentally, I released my lotto software in 1988. (Wow! How time flies! Two decades ago!) At that time, I was the only one who knew how to generate lotto combinations in lexicographical order. A few others were able to generate lotto combinations only in random manner. It took other programmers several years before they discovered the trick of lexicographical lotto generation. History always keeps accurate and unbiased records!

The lotto combination generation in lexicographical order is very simple today.

```
FOR A = 1 TO 49
FOR B = A + 1 TO 49
. . .
FOR F = E + 1 TO 49
```

The lotto generation in lexicographical order as *arrangements* is also very simple . . . to me!

```
FOR A = 1 TO 49
FOR B = 1 TO 49
IF B = A THEN GO TO 222
. . .
FOR F = 1 TO 49
IF F = E OR F = D OR F = C OR F = B OR F = A THEN
GO TO 666
. . .
666 NEXT F
. . .
222 NEXT B
111 NEXT A
```

Blackjack lexicographic hands can be only generated from a list, such as a deck of cards. Lotto combinations can be generated from a list as well. The list, however, must be ordered lexicographically (e.g., from 1 to 49, 49 elements, and 1 element per line). Blackjack hand generating faces a tremendous obstacle of its own. The number of elements per combination (or per arrangement) varies widely: from 2 cards to 10 or 11 cards per hand.

That was the challenge that kept gambling programmers paralyzed up until now. It is so much easier to say simulation! That is, generate blackjack hands randomly! Many kids can do that, like many adults. But I reserve the right to doubt everybody's accuracy in generating blackjack hands. I've seen so much falsity out there, including pathological lying and deception. Card counting is the most blatant example. Selling card-counting systems is still a sizable gambling business. All for naught!

This year of grace 2009 has proven to be augural to me. For I was able to successfully finalize the software to generate *all possible blackjack hands in lexicographical order*. It wasn't easy. I went through various methods and algorithms. Verification is the hardest part as there are no mathematical formulas regarding blackjack combinatorics and probabilities.

I generated blackjack hands as *combinations* or *arrangements*. Then I opened the output files (text format) and checked as many hands as possible. Yes, computing things are so much better today than just five years ago. The generating process is significantly faster. Also, opening large files is much easier today. My text editor of choice is my own *MDIEditor and Lotto WE*. It opens text files of several megabytes in size reasonably fast. The editor also uses a fixed-width font, which makes reading blackjack hands easier.

I started by creating one deck of cards. That is, write a simple text file with 52 lines, consisting of one card (number) per line, from the four 2s to the four Aces (written as 11). The file name is BjDeck1.TXT and is absolutely free to download. From that pivotal layout file (deck of cards), I deleted four of the 10s to get the deck used in the Double Attack Blackjack game. The new text file has only 48 entries (12 ten-value cards instead of 16). The file name is BjDeck2.TXT and is absolutely free to download. The output files for the arrangements generation are gigantic! Thus, I needed to create smaller deck files. The reduced deck files helped me tremendously with the verification process. One deck, BjDeck1-11.TXT, had only one suit of the thirteen cards (one 2, one 3, . . . four 10s, one Ace). The fourth deck, BjDeck4-11.TXT, had only one suit of the thirteen cards (one 2, one 3, . . . four 10s), except for Aces. All four suites of the Ace are in that layout file (sixteen entries in the file).

I created two more deck layout files (also completely free to download). One file was the result of shuffling (randomizing) the regular 52-line BjDeck1.TXT: BJDeck1Shuf.TXT. The other file was the result of reversing the order in the deck. The composition is from the four Aces to the four deuces: BjDeck1Rev.TXT. These two BJ deck files prove that the arrangements method of generating hands is the most precise. All three layout files generate the same amount of hands and the same bust percentage: BjDeck1, BjDeck1Shuf, BjDeck1Rev. By contrast, the combinations method of generating hands leads to three different result files.

1. The name of the first program is *BjDealerOrder.EXE*. It generates all possible blackjack hands as *arrangements*. That is, the order of the cards is of the essence. Total number of qualified (completed) BJ hands is staggering for all one-deck regular files: *334,490,044*. As for the *Double Attack Blackjack* game, there are *302,394,480* total possible completed hands.

I only calculated the hands but did not generate them. I simply commented out the statements for printing to an output file. I did generate, however, the arrangements for the BjDeck1-11.TXT and BjDeck4-11.TXT source files of incomplete decks of cards. The output files are also available absolutely freely as downloads: BjAllHands1-11Ord.TXT and BjAllHands4-11Ord.TXT.

Those two output files for incomplete decks played a role of biblical importance, as it were. They helped me discover the most subtle errors of generating blackjack hands of variable size (length). Generating hard-count hands was as easy as a breeze. The big problem came from the *Aces*, as they are counted either *1* or *11*. Blackjack hands such as 2+A+2+A must be hit by the dealer; hands such as 2+A+3+A are mandatory stand. I perfected my blackjack probability software due in great part to those two output files. One can easily check the error-free files in *MDIEditor and Lotto WE*. I have checked several times: each and every hand is absolutely correct, and no hand is missing. Of course, we follow the rules for the blackjack casino dealer. Yes, some casinos rule now that *hit soft 17* (e.g., A+6) is mandatory for the dealer. First off, that's a bad rule for the blackjack player; avoid such tables at all costs. Secondly, my software can be easily adapted to accommodate the *hit soft 17* rule.

2) The name of the second program is *BjDealerCombos.EXE*. It generates all possible blackjack hands as combinations. That is, the order of the cards is not important. Hands such as 2+A+2+A or A+A+2+2 are always written as 2+2+A+A. Total number of qualified (completed) BJ hands is far cry lower for all one-deck regular files: *297,615* (for 52 cards). As of the *Double Attack Blackjack* game, there is *257,877* total possible completed hands. I calculated the hands and also generated them. The output files are not available absolutely freely as downloads. They require the reasonable

Permanent Membership: BjAllHands1Combos.TXT and BjAllHands2Combos.TXT.

And now, the shocking results! If you go all the way down to the bottom of *BjAllHands1Combos.TXT*, you see that the bust percentage is *41.97%*. Wow! We all believed John Scarne and his "biblical" figure of *28%*! Keep this new figure in mind: *The odds for a blackjack dealer's bust are at least 41.97%.* The bust probability is calculated by dividing the number of dealer's busted hands to the total possible blackjack actions. *Blackjack action* is a parameter that counts everything: busted hands, pat hands (17 to 21), blackjack hands, and draws or hits (incomplete hands). The software does not print the *incomplete hands*.

The *best-case scenario* regarding dealer's bust probabilities for the game of blackjack reads,

Total BJ Actions: 476246
Total Draws (Hits): 178631
Total Nonbust Hands: 97735
Total Dealer BUSTS: 199880 (***41.97%***)
Total Complete BJ Hands: 297615
Total Blackjacks (10+A): 64 (4.83%)

The *worst-case scenario* is represented by the *arrangements* method of generating blackjack hands. The worst-case scenario regarding dealer's bust probabilities for the game of blackjack reads,

Total BJ Actions: 345948532
Total Draws (Hits): 11458488
Total Nonbust Hands: 110020276
Total Dealer BUSTS: 224469768 (***64.89%***)
Total Complete BJ Hands: 334490044
Total Blackjacks (10+A): 128 (4.83%)

I already experienced anger and hatred in blackjack and gambling forums. They would fight vociferously against this very theory of mine. They would not accept these truly mathematical figures for a number of reasons. For one, the vociferous advocates are strongly involved in the business of card-counting training for the game of blackjack. Some sell packages for upward of dozens of thousands of U.S. dollars. There are also the casinos. It is in the financially selfish interest of the casinos to advocate the application of blackjack basic strategy (BS) and, especially, card counting (BCC). The game of blackjack is highly profitable for the casinos mostly because of the BS and BCC.

And then there is that subhuman group of the afflicted. You surely know of UFO afflicts who are medically afflicted by irrationally believing in UFOs. You also know of religious afflicts who are medically afflicted by irrationally believing in gods (those saying, "God speaks to me.") The same is true about the milder and far smaller group of card-counting-afflicted blackjack players. Some of them are medically afflicted by irrationally believing in the true effectiveness of card counting. I can only feel for such individuals. But I am truthfully harsh against them pathologically liars who use lying and deception for financial gains. Pathologically deceiving liars who make statements on TV that they were abducted by aliens and tortured inside UFOs—and thus trying to cash in. Pathologically deceiving liars who broadcast TV religious programs, shamelessly claiming that God speaks to them constantly—and thus trying to cash in. Pathologically deceiving liars who shamelessly claim that they make big money by counting cards at blackjack—and thus trying to cash in. The latter three groups do it knowingly—hence, the pathological-lying attribute.

There will be also probability purists who will fight the blackjack figures my new blackjack software reveals.

They will argue against a number of blackjack hands generated by software: 2-2-2-2-3-3-3 or 3-3-3-2-2-2-2 or 2-2-2-2-3-3-11-11-11 or 2-2-2-2-3-11-11-11-11-3, etc. Such hands will never come out, they loudly proclaim! But hey, who decides what hands come out and what hands will never come out? Is there a god of blackjack who makes such decisions? NOT!

The purist argument resembles the older heated debate regarding the lotto combination *1-2-3-4-5-6*. Indeed, lotto combination *1-2-3-4-5-6* does come out so rarely that it has not been drawn in our lifetime! The standard deviation plays the determinant role. The same should be true about another random phenomenon as the game of blackjack. There are some issues here that we must address. Calculating the standard deviation for lotto is as easy as it can be (just use my *SUMS.EXE* standard deviation software). It is extremely hard to calculate standard deviation for blackjack output files. Think of those 10 GB files! We must have a 64-bit operating system and 64-bit compilers to create adequate software. Here is the most important issue. We know exactly how to calculate the probability of any lotto combination. For example, we can generate all 13,983,816 lotto 6/49 combinations in lexicographic order and see exactly one *1-2-3-4-5-6* combination. We must do the same thing in blackjack: calculate the probability precisely as $p = n / N$.

It took me a while to reach this moment in my activity. I am highly conscious of validity and validation. There is no doubt in my mind now that this new blackjack theory of mine is valid. I verified many times. Also, I made several verification/validation files available for free to everybody. The registered members of my Web site have access to additional witness files.

How about the software? My new blackjack software is **not** available for free or to the registered members of my site. The software will be available only via licensing. The license fee should include the source code, definitely. It will be available only by postal mail, registered and signed for confirmation. It will be on a CD without label, but showing my autograph. An agreement must be also printed on paper and signed. The most important action is filing for a *patent*—quite difficult for an individual without a lawyer. (Actually, there is only a *good* lawyer or ***no*** lawyer!) The *U.S. Patent and Trademark Office (USPTO)* already has in its library my software. They informed me that *USPTO* follows the rule of checking *any* patent application against their library—first and foremost. *USPTO* denies patent applications before potential legal action. My definitive decision will be announced on the *What's New* page at that well-known destination by now (*SALIU.COM*). You might want to check that page ever so often. I would **not** feel good if giving away such software to casinos or big-money gambling developers. They would exploit the players to even worse degrees.

I do have a *new blackjack strategy*. I haven't applied it yet. But it should work for me even better than before. Yes, I was successful before—just keep reading. For starters, figure out the BJ dealer bust probabilities this way: *41.97%* bust chance for the first hand at the table. Getting a second bust hand at the table is .4197 ^ 2 = *17.6%*. How about three busted hands in the same round? The probability would be .4197 ^ 3 = *7.4%*. And so on. See how well placed the blackjack dealer is? I know I may not place myself as well as the blackjack dealer. I can only place myself as well as the next position before the dealer. I don't know why they call it the third base—it should be called the *premier* base! How many players bust before me? The more of them bust, the more ecstatic I will be to buy them a drink or two or three. Problem is many casinos give the drinks for free, regardless of BS (basic strategy) or not.

2. The Fundamental Myth of Blackjack Gambling: Card Counting

You might have seen that movie ***21***. It had absolutely no success in theaters. A DVD was released in 2008 with much more success. The ***21*** DVD reopened the huge gambling appetite for the so-called surefire strategy of ***counting cards at blackjack***. The movie also introduced a powerfully symbolic ghost: the ***MIT Blackjack team***. If you watch all the features of the DVD, you will see the author of the original book that inspired the ***21*** movie. In his interview, the book author and the screenplay writer admitted that his book was the result of a rumor! How could people with the heads on their shoulders believe that an *MIT blackjack team* was even possible? MIT (Massachusetts Institute of Technology), such a prestigious institution, would even accept the rumor of a gambling team on the premises. Let alone a *real* gambling team consisting of faculty and student body! But adding ***MIT*** to a blackjack team did wonders!

The legend of card counting started with a well-written book, *Beat the Dealer!* The author, *Edward O. Thorp*, was a mathematician working for IBM. He also learned computer programming in order to prove his theory on blackjack card counting.

If the player keeps track of the cards that were dealt, there will be variable situations for the player. Thorp speculated that the situations were *favorable* to the player when *ten-valued cards and Aces (high cards) were predominant* in the remainder of the card deck. Reversely, the situation was unfavorable to the player when there were more small cards (2 to 6) compared to high cards. The 7-, 8-, and 9-valued cards were considered neutral.

The same John Scarne we talked about before puts *jokingly* the advantage of card counting. Suppose there is a one-deck blackjack game with 100% penetration (i.e., all cards are dealt). The player tracked the entire deck absolutely precisely. There are five cards remaining in the deck: three 8s and two 7s. The player would bet the maximum immediately (actually, millions if it were possible!) There is NO way the player can lose! The player would always stay on two cards (it doesn't matter if it is 7+7 or 7+8 or 8+8)! On the other hand, the dealer would always bust. It doesn't matter: 7+7 (under 17) draws an 8 and busts. Or 8+8 (still under 17) draws a 7 and busts. Or 7+8 = 15 (under 17); either 7 or 8 as the third card would bust the dealer's hand!

A situation like that would have occurred, but extremely rarely. To have two 7s and three 8s at the bottom of a 52-card deck has a degree of certainty in the same category as the moon colliding with the earth! Keep in mind, total possible permutations of 52 cards is calculated by factorial of 52 (52! = 1 x 2 x 3 x 4 x 5 . . . x 50 x 51 x 52). Who can say that number?

As a matter of fact, John Scarne challenged Edward O. Thorp to a real blackjack game in a casino. I quote from *Scarne's New Complete Guide to Gambling* (p.361):

> *In 1964, in an effort to test Professor Thorp's "winning Black Jack" statements I challenged him to a $100,000 contest to be staged in Las Vegas. Thorp's reply was a big "No".*

This excerpt is from page 348:

> *If he [Thorp] would like to team up with me and my partner to beat the Nevada Black Jack tables by*

> *making use of his unbeatable system. Thorp agreed and after the first three days of play in Reno, Nevada, we realized that Thorp knew nothing about the science of Black Jack play, and his countdown system never seemed to work . . . Thorp later admitted to us that he never really gambled.*

There is one more notable name to be mentioned here: *Ken Uston*. He became really famous when he appeared on CBS's newsmagazine *60 Minutes*. The curiosity of the TV news division was triggered by a successful lawsuit that Ken Uston had won. He charged that he was barred from playing blackjack because of his skills as a blackjack card counter. The court system decided that no player may be discriminated against based on the skills of the player.

The curious thing is that Ken Uston became a spokesperson for the casino he had filed lawsuit against: *Resorts*. He appeared in TV commercials aired in New York and other big cities of the Eastern board of the United States. Even more curiously, the *60 Minutes Ken Uston segment* was shot inside the same casino! Moreover, Ken Uston was allowed to win big money, as if it were a fictional movie! Meanwhile, if a regular casino patron tries to take a benign photo on the premises, he/she might as well be thrown in jail! Not to mention that *doing* card counting can result in harassment and eviction by force sometimes!

Card counting, as devised by Edward O. Thorp, is a footnote to gambling history now. It offered a slight advantage in one-deck games and especially toward the end of the deck. Ideally, a player could destroy the blackjack game *if* he knows the composition of the deck *and* the sequence of the remaining cards in the deck. The latter part is the real problem: nobody will ever be able to know the *sequence* of the remaining cards in the deck. The count may be +5, but often the sequence is *Low card, High card, Low, Low, High, High, Low, Low, High*, etc. The dealer has the same probability to get the high cards. It is even more complicated when one considers that there are several blackjack players at the table.

The cards, Low and High, will be distributed randomly among them. What makes a particular player believe he/she would be the one to get the High cards? In one-deck games, and only playing head-to-head against the dealer, there is a slightly higher chance for the player to get a blackjack. The dealer also has an equally higher chance to get a blackjack.
The difference is the player is paid 150% for a blackjack while the dealer always gets 1 to 1 for a natural. Let's say the player has a 2% higher probability to get a blackjack (for a positive count, including Aces).

Under normal circumstances (count = 0, neutral), the probability of a blackjack is just under 5%. What is 2% of 5%? That would be 0.02 x 0.05 = 0.001; that is, 0.1% better probability! That is, a 4.78% chance to get a blackjack compared to 4.77%. Applying the Fundamental Formula of Gambling to *p=4.77%* and *p=4.78%*, we get, approximately, the same median: 14.18 and 14.15 hands. On the average, there will be 14 or 15 hands between blackjacks, in neutral counts as well as in positive counts.

I have one problem with the two blackjack authors and researchers, Edward E. Thorp and Ken Uston. They reached a point of *worshipping* the method/system of card counting. I am sure they did a large number of computer simulations. I am sure they noticed the negligible effect of the system. Yet they continued to influence a large number of potential casino players that card counting represented the road to the riches.

I would like to present a few more excerpts on Thorp and Uston. *Carl Sifakis* writes in his 1990 *Encyclopedia of Gambling* (pp.36–37):

> *Dr. Thorp is still in computerized mathematical research, but he is now concerned with looking for values in stocks . . . The late Ken Uston, author of numerous books on counting, was at the end of his life involved in computer work in the Middle East, helping Kuwait track billions of dollars in investments. He was not playing blackjack in Atlantic City, although he had won a court case that barred casinos in New Jersey from refusing to let counters play. In fact, Uston, upon winning that case, didn't hit the blackjack tables in Atlantic City but signed up to do TV commercials for Resorts International, the very target of his suit . . . One long time gambler, Murray Friedkin, says of Thorp in "Big Julie of Vegas" by Edward Linn: "Thorp is the smartest man in the world; if you don't believe me, ask him . . . Whatever Thorp may say, I can tell you that if he has made any money on blackjack he made it by writing a book."*

Why this aura of legend surrounding card counting at blackjack? Even more mystique is added when considering that Las Vegas is still barring card counters from playing blackjack. Says Carl Sifakis, *"And what of the casinos today? Blackjack is a much bigger game today than it was before 'Beat the Dealer' appeared. More people than ever patronize the tables and casinos today make more money from the game than ever before. That's a significant bottom line."*

Roger Gros, senior editor of *Casino Player* magazine, writes in his 1996 *Casino Gambling the Ultimate Play-To-Win Guide* (p.30), *"After all, casinos make most of the money they make at the table games via blackjack. It's great advertising when someone reports a big win at the blackjack tables. Players have been encouraged for many years to believe that blackjack can be beaten, and the casinos don't want to do anything to disrupt that message."*

Indeed, the legend of blackjack card counting plays as the most successful means of advertising for the casinos. It certainly attracts a large number of players who believe counting at blackjack is a road to riches. There are also other ways that the card-counting legend favors the casinos. Read any blackjack book on card counting, and virtually all of them contain the same cliché: *"If you are a card counter, make some bonehead plays so that the pit boss won't ban you!"* I think many card counters take the advice seriously. They do make bonehead, stupid plays from time to time just to hide their card-counting skills. What a stupid strategy for the blackjack player! What a profitable play for the casino!

It is admitted that counting cards offers no more than a 2% advantage for the blackjack player. That's a slim margin by all standards. Making bonehead plays can easily wipe out the slim potential 2% advantage. The casinos owe big time to all authors of card-counting books. Then in places where it is legal to ban skilled players from the blackjack action, the casinos commit downright robbery.

From what I have read, the casinos show a strong bias toward barring a blackjack player when he/she is at a serious *loss*! I read once that a known card counter was losing some $14,000. Exactly at that point, the pit boss approached the counter and asked him to leave the blackjack game! Get it? The $14,000 went immediately to the casino bottom line. Meanwhile, the player was deprived of a reasonable chance to recuperate his loss. After all, blackjack is almost an even-odds game for a player using the basic strategy. It is fair to expect swings in the winning and losing columns of the player. Of course, the banned player is allowed to play blackjack again. The pit bosses pretend to have forgotten him/her until another significant loss for the player. The player is thrown out again! If the known counter is winning, the rationale is that a swing in the fortune will follow. Sooner or later, the player will encounter a severe loss. That's when you ask a counter to get out! What casino would be happy if a winner takes the money and run for real?

The *21* movie was about counting cards at blackjack. It glamorized the tactic by adding intelligence to the whole scheme: the *MIT* factor! The movie was shot inside the *MGM Grand* casino! Surprise, surprise! Would you read again the threatening message from the chairman of the same casino (2001)? He threatened my guests that the casinos would throw out casino patrons who would dare to write down roulette numbers! Yet the same casino allowed a big movie production inside their building! A movie production worshipping card counting.

3. The Theory of Streaks: Foundation of My Blackjack Gambling Strategy

Yes, I and casino blackjack have always had a glamorous affair. I admit, I was playing blackjack at the same tables with women who showed a strong interest in me at times. I preferred my affair with the game! No offense! A lot of mathematics is involved and implied in the game. Everything that has mathematics and potential to find more about Truth attracts me like pollen attracts a hardworking bee.

I was a farmhand since my arrival to the United States in 1985. It was hard physical work. I always looked at the positive side as well. The physical work balanced my intense intellectual work, including computer programming. Unfortunately, hard physical work leads to physical pain more often than not. In early spring of 1996, unbearable back pain took me out of farm labor. I was unable to find employment until late summer of 1998. I'm sure potential employers figured out that I quit my farm job due to some physical illness. Given the health care system in the United States, every employer has a major concern regarding job offerings. In any event, I recovered physically quite soon. But it was hard to convince employers that I was in good health physically. And that, above all, I had a special mind that could offer high benefits to the right employer.

All that turbulent period of searching reemployment, I kept working hard at developing theories and creating more and more software. Looking at the positive side, that period of unemployment made me a remarkable employer! I became my own employer. I was not tired anymore because of my hard work for others.

One of the things I did was study casino gambling. I used one of Thorp's terms: gambling mathematics. It was definitely mathematics. I saw also a means to make some money. I even thought numerous times of turning into a professional gambler. There was a problem nonetheless: bankroll. I barely had money just to survive. For one month, I only survived on food stamps from the government. Remember, the truth is above anybody and anything, including my pride. I am not ashamed of telling the truth.

I did receive employment in the late summer of 1998. I worked mostly in temporary jobs. Actually, I've been a temp ever since! Good thing I was paid. I was able to put food on the table and sleep under a roof. Additionally, I was able to save some money. I tried to build a bankroll.

Going to a casino was a freebie where I live. The casinos offered free bus drives. Actually, you paid $20, and the casinos gave you your money back on arrival. They are good at mathematics, aren't they? Just about everybody loses that $20 in minutes! Well, not me!

I hit the Atlantic City casinos a few times until the end of that year of grace 1998! I had a couple hundred dollars when I first started with blackjack. I returned home with more than the double of that amount. My window of opportunity was five hours. The bus trip would start early in the morning (Wednesday or Sunday). The trip to Atlantic City took around five hours. The trip back home would be five hours later.

The casino bosses were not mean during my first hits in 1998. First of all, I was a small-timer: no more than 200–300 dollars per session. Besides, they might have tried to figure out my system. It was clear for them that I applied a form of *martingale* betting. Most gamblers know that martingale betting requires doubling up the bet after each loss. Well, it was a whole lot more than that and a whole lot more refined in my case. The only negative reactions I recollect came from other gamblers.

Keep also in mind that my blackjack system was based on the wrong assumption that the dealer's bust at blackjack was 28%, as taken as a biblical figure in the gambling kingdom. As seen in this chapter, that figure is significantly higher. Therefore, the house advantage at blackjack is higher. Also importantly, the *player's position* at the table is *relevant*. At that time, I tried the middle position. By contrast, most blackjack players prefer the last position at the table (just before the dealer). *That is the best position at the blackjack table: the last seat or the third base. It is a* ***mathematical*** *law.* The more players before me bust, the lower the degree of certainty that I will bust—the dealer as well. It is worth repeating. The bust *probability* is constant while the *degree of certainty* of serial busts decreases. It is similar to consecutive heads/tails in coin tossing. Nobody gets three or four heads in a row with the same frequency as just one heads followed by one tails.

And thusly, we return to the concept of *streaks* once more. As I said before, probability theory might as well be named the *science of streaks*. The core of my casino blackjack strategy was founded on the application of streaks. The method required reasonably accurate records of the hands I played as *loss* or *win*. When I applied my gambling strategy in writing, I noted a loss by – (a minus sign) and a win by +.

Theory of probability, in general, is the *science of the streaks and skips*. A statement such as *"This event will always have the probability equal to zero point zero zero three four"* is virtually meaningless. The probability represents the ratio of the favorable cases over total possible cases. So it works with integers. In real life, we deal with integers (discrete values) such as numbers of elements and numbers of trials.

The events will *hit* or *miss* in *streaks* pretty clearly predicted by rules and formulas of probability theory. If you play longer sessions at the blackjack table, for example, you will face a higher probability of some very long losing streaks. But play shorter blackjack sessions, and there is a far better chance that you'll escape with shorter losing streaks. It is highly recommended to play blackjack more aggressively at the beginning of short sessions. Then move on to another table, preferably another casino.

A session at the blackjack table can be recorded as a succession of streaks. For example, from the player's perspective:

Win|Loss|WW|LLL|Push|L|W . . . etc.

The streaks can be also viewed as skips between hits (e.g., how many hands the player skipped between two hits [wins]). The underpinning of the streaks is undeniably mathematical. If you don't believe me, take Warren Weaver's word on it. Please reopen his informative book ***Lady Luck: The Theory of Probability*** at page 346.

The streaks can be calculated precisely, and real events concur with the formulas. Warren Weaver does the calculations by brute force. He takes the simplest and most fair game of chance: coin tossing. Let's say we consider a total of 1,024 tosses (it is 2 to the power of 10; easier to work with as the probability is 1/2). How many 4 consecutive runs of heads (H) can we toss in 1,024 trials?

Evidently, 4 heads should appear exactly in this pattern: THHHHT (where T is for tails). The combined probability of that pattern is 1/2 x {1/2 x 1/2 x 1/2 x 1/2} x 1/2 = 1/64. If we multiply 1/64 by the total number of tosses, the result is

1/64 x 1024 = 16 runs of 4 consecutive heads (or tails, for that matter).

I wrote software to calculate the number of streaks (or runs of consecutive outcomes) of various lengths for any given probability and number of trials. The program is *Streaks.exe*, and it is free for your enjoyment.

My presentation here starts with the old blackjack figure of a dealer bust equal to *28%*. Many blackjack players will still believe in that predicament for some time to come. I present here only the result of discarding the pushes. The approach considers the blackjack probabilities as follows:

~ dealer: p = 52% (or .52)
~ player: p = 48% (or .48)

Number of hands to play: 1,000 (or changing blackjack tables around 5 times a day)

~ dealer: 120 single-win streaks and 128 multiwin streaks
~ player: 130 single-win streaks and 118 multiwin streaks

Noticeable differences if playing 1,000 hands *continuously*. My software *Streaks.exe* and *Blackjack.EXE* always enforce the rule of playing fewer hands at the same table.

Following is the total streaks for the dealer and the player in 1,000 hands from single streaks (1) and 2–8 multiple like streaks:

Streaks	**1**	**2**	**3**	**4**	**5**	**6**	**7**	**8**	**Total Wins**
Dealer	120	62	32	17	9	5	2	1	505
Player	130	62	30	14	7	3	2	.8	475

The 20 extra hands are most mathematically likely allocated as 11 (52%) to the dealer and 9 (48%) to the player. Thus, the dealer can expect up to 11 consecutive wins in 1,000 hands—one time. The player can expect up to 9 consecutive wins in 1,000 blackjack hands—once.

The probability for the dealer to bust is 28%. In a majority of cases, the blackjack dealer busts in 2 consecutive hands or skipping 1 hand or skipping 2 hands. The same is *not* true for the player. The basic strategy blackjack player busts 17% of the time. Therefore, the streaks of the player's bust are longer.

Let's calculate the number of streaks of losing exactly *four consecutive blackjack hands* (from the player's perspective). We symbolize a dealer's win by **L** and a player's win by **W**. We can calculate the number of the streaks consisting of exactly four consecutive losses for the player.

That is, calculate the number of the streaks in the format **WLLLLW**. If probability for W = .48 and the probability for L

= 52%, then the combined probability is .48 * (.52 * .52 * .52 * .52) * .48 = .016845963 or *1 in 59 or 60* hands. Mathematics of gambling expects such a streak once every 60 hands or twice in 120 blackjack hands. Or 17 times in 1,000 continuous blackjack hands. Yes, it is possible, albeit mathematically rare, that the blackjack player can experience 3+ such streaks in 100 hands. Equivalently, however (casino executives hate the equivalent thing), the blackjack player could experience 100-hand sessions with all losing streaks shorter than 4 (3-)! Those are likely sessions with longer player's win streaks, usually with more frequent dealer's busts!

Now, the more pessimistic reality in blackjack, from the player's perspective. I am a strong believer in the validation of any idea or theory. My latest blackjack software (2009) undeniably proves that the percentage of the bust hands is, at least, ***41.97%***. The bust hands were generated following the dealer's rule: *Hit all hands under 16 and stand on all hands 17 or higher.* In this case, the house advantage (HA) at blackjack becomes ***41.97% x 41.97% = 17.6%.*** That HA figure is valid only for one player against the dealer. If we deduct the traditional 4% gained by the *basic strategy player*, we reach a ***14%*** house advantage at blackjack. Again, I discard the pushes (ties). The streaks will be worse for the player under these new circumstances (which I believe to be the correct ones). No wonder so many blackjack players are dismayed to hear that blackjack is almost a fifty-fifty game if playing basic strategy! The players lose far more than the so-called mathematical expectation. I did hear casino players at the blackjack tables expressing anger after quickly losing all their chips. It was better to play roulette, some said. I certainly agree with them right now. *The house edge at roulette is better than in blackjack!* Yes, some will always stigmatize you: *roulette is a sucker's game!*

The ***41.97%-dealer-bust*** approach considers the blackjack probabilities as follows:

~ dealer: p = **57%** (or 0.57)
~ player: p = **43%** (or 0.43)

The number of hands to play is the same: 1,000 (or changing blackjack tables around 5 times a day)

~ dealer: 105 single-win streaks and **139** multiwin streaks
~ player: **140** single-win streaks and 105 multiwin streaks

Streaks	**1**	**2**	**3**	**4**	**5**	**6**	**7**	**8**	**9**	**10**	**Total Wins**
Dealer	105	60	34	20	11	6	4	2	1	1	561
Player	140	60	26	11	5	2	1	-	-	-	420

The 19 extra hands are most mathematically likely allocated as 11 (57%) to the dealer and 9 (43%) to the player. Thus, the dealer can expect up to 11 consecutive wins in 1,000 hands—one time. The player can expect up to 9 consecutive wins in 1,000 blackjack hands—once.

There is a **significant** difference that favors the blackjack dealer—that is, the casinos. A number of gamblers cautioned me with regard to John Scarne. Indeed, he helped the U.S. government with gambling matters, and especially defense against cheating the U.S. troops by gambling providers. On the other hand, he had a serious financial interest in the gambling industry. He was friends with Frank Sinatra who, as loudly rumored, had friendship in the mob world. Perhaps John Scarne *intentionally* published wrong calculations pertinent to the house advantage at blackjack! The *real* odds would have scared away players from the most attractive and most popular casino game! Who knows?

But what *I know* for sure is that the new data that I calculated is in tune with the reality. I was like most gamblers. I suspected dealer cheating quite a few times. Why does that dealer win so many hands in a row while my wins come in single streaks that frequently? Believe me, I looked at dealers' hands so intensely sometimes that I risked being treated as hostile! I have never observed foul play, however. I played for fun in a Detroit casino in the summer of 2009. I spent some time with my daughter. One of my main points was the *length of the streaks*. Why did the dealer have so many long winning streaks (against me) while my winnings came mostly in the meager *one hand a time*? I remember asking my daughter if she remembered my streaks of 5 or longer hands? She was keeping track as well. I did not have a 5-streak in about 100 hands! As you see in the table above, a 5-streak of consecutive winning hands occurs 5 times in 1,000 hands. That is, less than one occurrence (.5) in 100 hands.

This is what led me to the intense activity of writing new blackjack software to generate all the hands and every correct bust hand. *Trust me and verify me!*

The following passages show how I applied my streak strategy at the time when I thought the dealer bust was **28%**. I also considered that player's probability was 48% while dealer's was set to 52%. In the light of the new blackjack odds, the situation is worse for the player: 43% to 57%. *In hindsight, my previous blackjack strategy was more aggressive than mathematically advised.*

Yet I was able to win consistently up to 2003 when I reached a serious bump on the road. The floor manager of an Atlantic City casino prevented me from playing blackjack. I had been impeded to play roulette the same day in the same casino. I did carry a pocket notebook (pencil-and-paper) and wrote down my losing and winning streaks. It was not a cheating device by any stretch of imagination! A pencil is not an electronic device like a miniature camera to peek at the cards in the deck! I never touched a deck of cards or a roulette wheel. I documented thoroughly the incidents and published several pages at my Web site: http://saliu.com/winning.html. You will find exact times and badge numbers and even names of the casino personnel involved.

I didn't use a notebook in my previous casino outings. I memorized my streaks in this manner:

My count always starts at *0 / 0*. I say mentally *zero-zero*.

~ The first number represents my *consecutive losses.*
~ The second number represents *my skips without two or more consecutive wins*. This parameter is trickier to grasp. I still have difficulties keeping accurate track of it mentally. It is much easier on paper.

If I *lose* my first hand, the count becomes *1 / 0*. If I lose also the second hand, the count becomes *2 / 0*. I say mentally *two-zero.*

If I *win* my first hand, the count becomes *0 / 0 and **a half***. I say mentally *zero-zero and **a half***. The ***half*** simply indicates that the parameter is still running. If I lose the following hand, the count becomes *1 / 1*. I say mentally *one-one*. That means I lost 1 consecutive hand; I also have a streak of 1 without winning 2 hands in a row.

If I *win* my first hand, the count becomes *0 / 0 and a half.* I say mentally *zero-zero and a half.* If I win also the following hand, the count becomes *0 / 0*. I say mentally *zero-zero*. That means I have no consecutive losses; I also won at least 2 hands in a row.

If I *win* my first hand, the count becomes *0 / 0 and a half.* I say mentally *zero-zero and a half.* If I *lose* the following hand, the count becomes *1 / 1*. I say mentally *one-one*. That means I lost 1 consecutive hand; I also have a streak of 1 without winning 2 hands in a row. I *lose* also the following hand. The count becomes *2 / 1*. I say mentally *two-one*. I *win* the following hand. The count becomes *0 / 1 and a half.* I say mentally *zero-one and a half.* I *lose* the following hand. The count becomes *1 / 2*. I say mentally *one-two*. It means I lost 1 consecutive hand; I also have a streak of 2 without winning 2 hands in a row. That is, in 2 consecutive situations, my win was not followed immediately by another win.

I noticed that the second count is less streaky than the first one. That is, I win more regularly 2 (or more) hands in a row. I keep that in mind when I martingale my streaks. The Fundamental Formula of Gambling tells me that I win at least once in 7 hands with a degree of certainty of 99%. I martingale more aggressively at the beginning of the game. I martingale the first count when it reaches 4. Since the second parameter is more consistent, I martingale it after it reaches 3. I go up to 7 in both cases (2-4-8 and 2-4-8-16). I stop at 7. I become more cautious after 100 hands or so. I martingale the first count after it reaches 6 (or 5 earlier in the game) and the second count after it reaches 5 (or 4 earlier in the game). I go as much as I can or allowed. If I sense that the streak is real bad, especially inside the same shoe, I stop the martingale after 7 and start another one with 3 bet units.

The advantage at blackjack is higher payouts for blackjacks and double downs. After long losing streaks, the winnings come quite often as blackjacks or double down hands. I also try to disturb long streaks that are favorable to the dealer. There are shoes when the dealer seems to always have blackjack or 10-10 or never-busting hands. Some players play an extra betting box, if available. Others sit out 1 hand. Personally, I try not to play by the book. If playing by the book helped the dealer in this shoe, I figure I'll break the rules this hand and thus hope to break dealer's hot streak. As I said before, I split two tens against dealer's 4, 5, or 6. I also hit 12 against 4, 5, 6 in those situations. Or I stand on low count against dealer's 7-A. Unfortunately, some players get mad at me when I do that. Point is it's my money, and I play the way I want to.

In the *write-down version*, I track also the bust hands, both for dealer and for my hands. The clear majority of the bust situations are *repeats* after 0 hands, 1 hand, 2 hands (0 hands means *consecutive* occurrences). If the dealer bust busted, I expect her to bust again in the next hand or the following one or after two more hands.

A ruled pocket notebook is the best format, especially if the number of lines is a multiple of 10 (e.g., 20 lines per page). That makes counting the streaks much easier. I was always quick with my counts and placing my bets. I was never at fault for the flow of the game. I did not impede any other player to play their game.

I recreate here a fragment with my real play at Trump Plaza, July 21, 2003, beginning 5:30 p.m. The blackjack dealer was Ginny.

Line number	Hand parameters	Hand parameters
1, 11	10-	20+
2, 12	10+	50+ B
3, 13	10-	20-
4, 14	10-	25- b
5, 15	10-	50+
6, 16	0	10-
7, 17	30+ j	10- j
8, 18	20-	25+
9, 19	10+	25+
10, 20	40+ B	10+

The interpretation.

As you can see, I did not go for the kill: my maximum bet was $50. The minimum bet was $10, the maximum limit at the table was $500. The *1 to 100* ratio of the bet spread is the minimum I accept. The + sign means I won (e.g., *40+ B* [line no. 10] means I bet and won $40; the dealer busted that hand). The minus sign means I lost that hand (e.g., *25—b* [line no. 14] means I bet and lost $25 while busting). I always differentiate between a dealer's bust (B) and my busted hand (b). The *j* means a *blackjack* or *natural* (both for dealer or player). The session was short (under 100 hands). My longest losing streak was 5 consecutive hands. My longest winning streak was 4 consecutive hands (2 occurrences). I won over $300, not counting my generous tips. I strongly advise now against tipping the casino dealers. The casinos already have the odds in their favor, not to mention the house edge. The owners and the executives should cut their fat incomes and pay better the little guy (who takes a lot of heat to enrich the bosses).

Yes, things can go wrong even in short sessions. A player might encounter a really long losing streak. As a matter of fact, that's the mantra of all casino executives or gambling gurus who have a strong financial interest (selling nonmathematical systems). They will tell you that you are going to lose in very long streaks. But they never mention the opposite: the player can win in very long streaks in short sessions. Things happen both ways. They are mathematical, but they represent the exception, not the norm. If you experience a long winning streak, my advice is to leave that table as soon as the hot streak ends.

Baccarat

I never liked that casino game. It is cold and strange—I mean minibaccarat. I never took a seat at the stupidly aristocratic game of baccarat! The players at the table look at one another, waiting for the first move (bet). Some players skip several hands in a row.

I am even more reluctant now. I am not sure if the (mini) baccarat odds are calculated correctly. The odds, once again, are based on computer simulation. We know that simulation skips quite a few legitimate hands, according to *Ion Saliu's Paradox*. Nobody knows what hands are in and what hands are left out. I don't know when or if I'll have the inclination to write baccarat software to generate all possible hands.

As of gambling strategy, one should apply the same system I applied to blackjack. The advantage is that they allow writing down the results of the baccarat hands. They give you pencil and paper, but I strongly recommend the same type of ruled pocket-size notebook. The number of streaks is very similar to the first table of blackjack streaks in this section (28% bust odds).

Craps

This is one of the most popular casino games in the United States. The odds of the game are much, much easier to calculate: 2 dice, 12 point faces. A few craps bets are very close to a fifty-fifty game. My blackjack strategy can be applied, but more aggressively. It is hard to write down at the craps table as there are no seats. Only the mental version can be applied. The advantage of writing: the player can track more than one type of bets.

Chapter XII

THEORY OF PROBABILITY APPLIED TO ROULETTE

1. My Early Roulette Strategies and Systems

I had seen dozens of roulette systems advertised all over the Internet. The reaction of the public was largely negative. From what I understood, the roulette systems had no mathematical basis whatsoever. Like one roulette system based on always playing 6 magical numbers no matter what. Most of the roulette systems were terribly expensive, even more so than the card-counting systems for blackjack.

Enter Ion Saliu. I applied my betting strategy as described in the blackjack chapter. I decided to sell one simplified system in 2000. Selling gambling systems is a very delicate action. It could be dangerous physically. The price: a shockingly low $50. Compare that to one roulette system advertised for $40,000! You can read a whole lot more, again, at my Web site.

I did receive ecstatic reactions from a few users of my roulette system. He even e-mailed to me and wanted me to publish his message on my message board. The hostile reaction, however, was a thousandfold more intense. Too many headaches for just $50—I withdrew the offer. It is easy to figure out the reaction. There are also customers who realize how good a thing is. They don't want many other people knowing about it. Sometimes thickheaded reactions can influence a seller to give up. I waited for some learning in the legal field so that I would come up with binding contracts.

Furthermore, you already read the threatening remarks of a casino chairman in retaliation to my roulette systems.

Yes, I felt the need to retaliate! Why not me? I offered a totally free roulette system that is known today all over the world: ***Free #1***. I launched it as *Handy-Brandy Roulette System*. Indeed, the system appears so scary that it should be played with a glass of brandy in one's hand! You can guess the intensity of the reaction, both on the positive side and especially on the negative side!

The ***Free Roulette System #1*** was derived from what later on became known as *Ion Saliu's Paradox*. If one plays all 38 numbers (double-zero roulette), the probability to win is 38/38 = 1. That is, full certainty. But that is a *tautology* (in logic) or an *absurdity* (in mathematics). Such a bet makes no sense, as the player automatically loses 5.25% (the house edge). The player is only paid $36.

If the player bets one roulette number at a time over 38 spins, the winning probability is 63.7%.

If the player bets 34 numbers in one spin, the degree of certainty goes up to 89.5%.

You can see the gain or loss for the three betting situations in the following table.

Type of Play	Winning Chance	Cost	Gain %
1 # 38 spins	63.7%	38 units	-
38 #s 1 spin	100%	38 "	36.3%
34 #s 1 spin	89.5%	34 "	28.8%

A so-called *wise gambler* is more than happy to play one number at a time. What he does is simply *losing slowly*! Not only that, but losing slowly is accompanied by losing more. That cautious type of gambling is like a placebo. A roulette system such as ***Free System #1*** scares most roulette gamblers. *"Play 34 or 33 numbers in one shot? I'll have a heart attack!"* In reality, the ***Free Roulette System #1*** offers a ***28.8%*** advantage over playing singular numbers in long sessions. In fact, the gain is higher since the bets are placed at optimal moments. That's mathematics, and there is no heart to worry about. But brandy might be still needed! How to play it?

A diligent, disciplined player can make a weekly profit by *covering all the numbers but four, playing no more than 10 spins a day*. The four numbers discarded are the last four spins. You wait patiently, not playing, until the following event occurs. *When one of the last four numbers hits again, you get ready to play.* You cover all the numbers (including 0 and 00) but the numbers in the last *five* spins (four numbers to play because one is a repeat). *You place 34 2-unit straight-up bets.* Chances are you will win most of the time. Next roulette spin, you make the same bet using now three units per number (you are using some casino money). The first time, you won 38 – 36 = 2 units; 2 units times 2 = 4 units in profit. The second time: you won 2 units x 3 = 6 units. Total profit per two plays: 10 units. Things are not always like this. Handy-Brandy could lose sometimes the first bet. It happens very rarely, but one of the last four roulette numbers will hit again in consecutive sequence.

In most cases, Handy-Brandy wins two times in a row without any of the last four roulette numbers repeating. He goes immediately to another table. He knows that the long run destroys the casino roulette player more than anything else. After winning at the second table too, Handy-Brandy usually cashes out. He runs out of the casino. Sometimes, he might try doing the same things at the third table, but never more than winning at three roulette tables. The player will do the same thing the next day at a different casino and again at another casino.

Here is illustrative data from the Hamburg casino. The following figures do not represent roulette numbers, but *skips*. The skips, as in the frequency reports above, represent the number of spins between hits.

9 5 57 17 30 18 ***0*** 3 13 25 6 ***0*** 4 17 2 24 ***0*** 4 3 15 13 8 21 2 23 2 6 7 7 10 3 3 17 10 15 11 1 15 3 1 24 11 10 2 3 13 6 12 3 3 24 6 10 21 4 8 9 8 ***0*** 20 7 21 7 18 6 ***0*** 17 4 13 5 2 6 16 ***0*** 2 4 ***0*** 7 ***0*** 7 1 8

The entry points are marked by a bold, italicized ***0***.

A figure, such as 57, represents the skip of a particular roulette number. That particular roulette number (the third most recent in our database) hit after 57 spins. That is, if we look at index 57+3 in our database, we'll find the same number again. Zero in the string above does not represent the *green 0* (roulette number 0). It means that a particular number hit two spins in row or hit in two consecutive spins. It couldn't be clearer now.

The free roulette software program, *Spins.exe* (also *Roulette. EXE* and especially *Roulette.EXE*), offers the best graphical representation of this free system. Scroll down to the end of the frequency report. The system of interest is named *Play all the numbers except for the last 5 spins*. The report also recommends playing immediately following two consecutive misses (indicated by the – sign).

You only play at the points marked ***0*** (zero). That's when the last number is a repeat from the previous 5 roulette spins. You will play, as above, the next 3 points where the skip is zero. In many cases, it's after the third win in a row when you can encounter a string like ***0 0***. When that happens, you will *lose 34 x 2 = 68 units*. You improve your chances further when you leave after you win at three roulette tables. You should expect to lose once a week.

2. The Wheel Bias: Roulette Strategy Based on Wheel Half (Hemisphere)

There is a lot of talk about ***roulette wheel bias***. That is, certain numbers or groups of neighboring numbers tend to appear more frequently on the same roulette wheel. Different roulette wheels favor different groups of numbers. If an observant gambler detects the bias, he or she is bound to big riches.

One *great* device to detect bias has always been a so-called *roulette computer*. Such hidden device is allegedly capable of *clocking* the speed of the wheel. The same speed spins the same roulette numbers (or small groups of neighboring numbers). So well has such device been hidden that nobody, but nobody, has ever seen one!

Clocking the speed of the roulette wheel is an illusion the myths are made of. NASA's sophisticated instruments may be able to measure precisely the speed of the roulette wheel. But no instrument can predict the landing position of the spinning ball on the roulette wheel. The wheel deflectors randomize the movement regardless of the speed. Of course, it is absolutely impossible for any roulette player to clock the roulette wheel speed precisely, even if the casinos allowed the use of any instruments/computers on the premises.

Not to mention the thorough tests conducted by the manufacturers of roulette wheels. They certainly perform tests regarding the wheel speed and the landing of the spinning roulette ball. They probably run the roulette wheel at exactly the speed, v (rotations per minute), for a number of runs, R. They note that the roulette ball is landing randomly (i.e., in various positions on the wheel). They change the wheel speed and run another batch of tests. They notice the randomness of the ball landing. It is a strong fact regarding the roulette wheel. The same speed of the roulette wheel leads to highly random landing positions (roulette numbers).

The ***bias*** exists, however. I did notice it myself. I detected serious roulette wheel bias in two very different situations: two different casinos in two different countries.

The first bias I discovered was due to the wear of the roulette wheel. Probably the deflectors of the roulette wheel were worn out. The wheel was malfunctioning; anyhow, that's what I thought.

We can divide the roulette wheel in two sections based on the last number drawn. There are 38 numbers at double-zero roulette. The last number drawn, plus 9 numbers to its left on the roulette wheel layout and 9 numbers to the right on the layout, make up the *inside hemisphere* (*half*) of the roulette wheel. The remaining 19 numbers make up the *outer hemisphere*.

The next number should be equally distributed, based on the FFG, between the two wheel hemispheres (halves). That is, the roulette *insiders* and roulette *outsiders* should be FFG-equally distributed. I tracked the spins at a roulette table in Atlantic City in July of 2000. To my (belated) dismay, the ratio was 80%–20% *biased* toward the *outer* hemisphere! Actually, two roulette dealers—a woman and a man—shifted at the table during that continuous run! So it was not a so-called *dealer's signature*! It was a malfunctioning of the roulette wheel. It is likely that the casinos have no knowledge of such phenomenon. I only detected the bias after my return home after looking at my notebook. I improvised a piece of software quickly, and I was really shocked to see that unexpected, but strong, bias.

Tracking the landing in the roulette inner hemisphere and the outer hemisphere is very hard to accomplish when actually playing roulette in a casino. I still decided to write software to run a roulette system based on the *wheel hemispheres* or *halves*. The name of the roulette program: *RouletteHemis.EXE*. I first offered my software as a gift to a tough foe in 2002! Why? Because he had given me a good piece of information regarding another individual who claimed he was Ion Saliu and was selling my gambling system in an Atlantic City casino!

I discovered that my roulette hemisphere system was pirated soon after my generous gesture! The climax of piracy was reached in 2009. An Australian group behind the Web site named www.genuinewinner.com already sells a $2,500 system. They try their hardest to avoid specifics of their roulette system. But one picture is worth a thousand words. Their advertising logo reads (in an image),

> *We consider which HALF of the wheel the ball landed in.*

They, the pirates, do *not* offer software to be downloaded to the player's computer. They might be using my *RouletteHemis*.

EXE behind the scenes! They also warn against playing roulette at online casinos because the wheels are not mechanical! They claim millions of dollars in wins, especially in the United States! I believe I right a wrong by offering to the public at-large the same piece of software—and then some. The original system is much improved. I added a second grand system based on the so-called *birthday paradox* (probability of coincidences or duplication).

August 26, 2009—the Australian roulette pirates got really scared by the law! They announce at their Web site that they decided to stop selling their roulette system and roulette computers. They already sent e-mails to their customers on the decision to cease selling the pirated version of one of my roulette systems. Guess how the deceived roulette customers will react!

The double-zero roulette wheel can be divided in two equal hemispheres of 19 numbers each. The single-zero roulette has uneven “hemispheres” or halves: 19 numbers for *inner* and 18 numbers for *outer*.

I had tracked only 40 spins at the same Taj Mahal roulette table on July 2, 2000. Indeed, the data sample seems to be meager. Keep in mind, however, that 40 is greater than N for the game of roulette. Comparably, N for a *6-from-49 lotto* game is almost 14 million! In my roulette sampling, the *outer* half recorded 32 hits, with only 8 hits *inside*! A *4-40-1* disproportion (bias) is a strong indicator of serious malfunctioning of the wheel.

I do have a whole lot larger data sample for 0-roulette. I collected almost 8,000 roulette spins at the same table at Hamburg Spielbank (casino) in Germany. The widespread consideration is that Hamburg Spielbank offered accurate roulette results. They stopped offering data quite often, especially after I published a gambling page or two at my site. They cancelled data publication for table 31 where I collected and published every roulette spins for the entire month of January 2006. You might never find roulette *permanenzen* (actuals) again. Or you might find from time to time—but only from very few casinos.

The Hamburg Spielbank roulette is single zero: a total of 37 numbers from 0 to 36. The inner half consists of 9 + 9 + 1 = 19 numbers. The outer hemisphere (semicircle) consists of the rest of the roulette numbers: 18. Thus, the ratio between the outer half and the inner half is *18/19* = ***94.7%***. The statistical results should show a comparable ratio between the two halves (or semicircles or hemispheres).

The analysis of the last 1,000 spins (most recent) of the Hamburg Spielbank database shows an outside/inside ratio of *93.5%*. The bias is 94.7 – 93.5 = *1.2%* in favor of the inside half. The analysis of all 7,990 spins of the Hamburg Spielbank database shows an outside/inside ratio of *90.6%*. The bias is 94.7 – 90.6 = *4.1%* in favor of the inside half.

Apparently, the roulette table no. 1 at Hamburg casino had a run when the inner hemisphere was statistically biased. The 4.1% bias is an average for the 7,990 spins in the data file of Hamburg Spielbank permanenzen. If 1.2% was a minimum, then a maximum of over 7% occurred in any 1,000 groups of contiguous spins. It is possible that maximum biases larger than 7% occurred in runs shorter than 1,000 roulette spins (e.g., 200 spins). A new version of my roulette software will allow for statistical analyses within ranges of spins at a user's choice. Right now, the analysis starts at the top, always spin no. 1 (the most recent one). The last spin to be analyzed is determined by the user, and it can go all the way to the bottom (the oldest result in the data file).

Standard deviation is the watchdog of randomness. For 7,990 spins, the standard deviation for the wheel outer half is 45 spins. That represents 0.6% of total spins. That is, the fluctuation should be plus or minus .6%. It is the overwhelming case of unbiased roulette wheels. The 4.1% bias falls way outside *three standard deviations* (*1.8%*)!

For 1,000 spins, the standard deviation for the wheel outer half is 16 spins. That represents 1.6% of total spins. That is, the fluctuation should be plus or minus 1.6%. It is the overwhelming case of unbiased roulette wheels. The 1.2% bias falls within three standard deviations. The roulette wheel bias cases of over 7% do not. They fall way, way outside the normal probability rule!

The case of three standard deviations is really abused and misinterpreted. Take the 1,000-spin case. If we analyze 1,000 ranges of 1,000 spins each, only in one case will we see a situation where the bias is close to 3.6% (three times 1.2%). In most cases, the roulette bias will be much less than 3.6%. Keep in perspective the rule of normal probability!

I did some calculations for one month, that January of 2006. Playing this roulette wheel half system with flat betting would have yielded a profit in excess of a quarter million euros. Had the player become aware of the inner-half bias of roulette table no. 1 at Hamburg Spielbank (by just analyzing the results on the table marquee on January 1), he/she would have netted well over one million euros by applying a 3-4-step martingale betting!

Indeed, millions and millions would have been profited from exploiting the roulette wheel bias. Chances are the casinos had no clue regarding this type of roulette wheel bias—before I released my always-unique software. And thusly, another gist of pirates tried to exploit Ion Saliu's intelligence for their own profit. Main point is they committed piracy—that's against all laws in all countries. More sadly, the pirates charged big money for a gambling system and roulette software program already available for free! I mean, just pay a modest U.S. $29.99 membership free—and everything else after that is totally free to use! Don't have to pay U.S. $2,500 for something that you already get for free!

Most likely, the Hamburg casino did not change the wheel during that month. They were not aware of this type of statistical analysis. It was not easy for a potential gambler either. She would have had to be in the casino all the time, at the same table. Looked at the marquee and quickly realized that the inner half was favored. Started betting on numbers in the inner hemisphere. Applied progression betting after a loss (when the roulette ball landed in the outside half). If the player was lucky to find herself in the middle of a strong bias (like I was in that 20/80 case), she would have collected big money, even millions! My case was for double-zero roulette; therefore, the ratio between the inner and outer halves is even: 19/19 = 100%.

The latest reports indicate that the casinos change the wheels quite often! The casinos also change the wheel layouts or patterns more frequently than ever. The player has to be really quick at reading those table layouts (cards) that the casinos offer in their roulette areas.

Following is a fragment from that biased data file I detected at the Hamburg Spielbank roulette table no. 1 in January 2006.

Spin #	Number Drawn	Distance in Slots	Inside – Outside +
1	11	7	-
2	18	14	+
3	32	10	+
4	22	16	+
5	6	12	+
6	31	10	+
7	4	11	+
8	30	3	-
9	27	6	-
10	21	7	-
11	14	2	-
12	16	14	+

The **Distance in Slots** takes values between 0 and 19. If a number repeats, the **Distance in Slots** is 0. The number of slots is counted either to the left or to the right of the previous number drawn. If the number of slots between two consecutive numbers drawn is larger than 19, then the counting is done in the opposite direction. For example, the number of slots between number 0 and number 16 is 21—if counting from right to the left; if counting from left to right, there are 16 slots between 0 and 16. We always record the lowest number of slots. We can also see that 21 + 16 = 37: total numbers on the single-zero roulette wheel.

The player detected bias toward the ***outer*** hemisphere, so bet on it. At spin no. 1, the player lost. Started a martingale at spin no. 3, doubled up and won. Then one more time at spin no. 4. The player kept betting the minimum on the ***outer*** wheel half—and won 6 times. There was a loss at spin no. 8. Doubled up spin no. 9, then no. 10, and no. 11. Won at spin no. 12 after a three-step martingale bet. The ***outer*** hemisphere consisted of 18 numbers. The player won 9 spins as spins no. 9, no.

10, and no. 11 had losses that were eventually recuperated at no. 12. The straight-up bet is 36 or a net of 16 units. The net profit would be 9 x 18 = 162 bet units in around two hours of play.

3. The Wheel Bias: Roulette Strategy Based on Number Pairing

If there is such a thing as the wheel bias, then a few roulette numbers will be drawn significantly more often than the other numbers. Also, each roulette number will be immediately followed by a limited universe of other roulette numbers. The latter parameter is known at my Web site as *number pairing*. It is common in lottery. You can read there much more about the *wonder grid* in lotto games. I will also write a few facts in this book in the chapters dedicated to lottery.

A *pairing* in roulette is a roulette number *and* the number drawn in the very next spin. In the case of a consecutive hit, a roulette number is followed by itself. Let it be said that each roulette number can be followed by itself with an equal frequency to a pairing consisting of a different number. The probability is the same.

How about the degree of certainty? I noticed that a repeat is less frequent than a heterogeneous pairing. I generated two statistical reports based on real casino roulette spins. I analyzed 1,968 real spins recorded at the Hamburg, Germany, casino, February 1–6, 2000. File name: HAMB00.DAT freeware from my download site.

I went back online to the Hamburg casino (Spielbank) in Germany (or *Cyber Germany*, every country of the world has now a physical location and a *virtual* or *cyber* location).

I downloaded the results for one table or wheel (no. 1) for the entire month of January 2006. File name: HAMB0106.WH1. It has 7,990 lines (draws or spins) in text format; the last draw on January 31, 2006, is at the very top of the file (line no. 1). The file contains roulette numbers (spins) only, one number per line. The casino result files have also other data such as frequency and various statistics. The files also have dashed lines (—), which represent dealer changes.

HAMB0106.WH1 represents the best recording format: table by table. Do not mix the spins from different roulette wheels in the same file. We would be talking very different beasts!

The *roulette pairing reports* can be viewed in their entirety at my Web site. The reports are quite long! You can also generate new reports based on your data.

The first report is for the most recent 1,000 spins: approximately 4 days worth of spinning or from January 27 to 31, 2006. The second report analyzes all 7,990 roulette spins for the month of January 2006. I do not see any severe bias according to the normal probability rule.

There is a number of 54 consecutive repeats in 1,968 spins. That amounts to about 27 roulette numbers in 1,000 spins. There are 37 roulette numbers in the single-zero game (as in Hamburg, Germany). It means that a few roulette numbers did not hit in consecutive spins for the duration of a 1,000-spin session. But each roulette number was followed in the very next spin by a different number at least 2 times in 1,000 spins.

For example, the roulette number *0* was followed by itself (*0* again) in the very next draw (spin) once in 1,968 spins. Roulette number *0* was followed by roulette number *1* in the very next spin 6 times in 1,968 spins.

Roulette number 1 was followed by itself (1 again) in the very next draw (spin) once in 1,968 spins. Roulette number 1 was followed by roulette number 29 in the very next spin 5 times in 1,968 spins.

Roulette number 5 was followed by itself (5 again) in the very next spin 0 times in 1,968 spins. Roulette number 5 was followed by roulette no. 21 in the very next spin 4 times in 1,968 spins.

Even if there are clear discrepancies, they fall within *3 standard deviations from the norm*. Therefore, the statistical data is still compliant with the rule of the normal probability. The discrepancy should have been two times wider to turn into suspicious bias (e.g., from 1 to 12).

A second take is in tall order: *mechanics*. There should not be any discrepancy between a *mechanical roulette wheel* (as in the casinos) and a *random number generator* (as in roulette software). The bias against an immediate roulette number repeat is less pronounced in random generation of roulette spins. A short explanation: the *croupier* (dealer) places the roulette ball in the slot that hit last. The probability is lower to have the roulette ball come exactly in the same slot as in the previous spin. We are dealing here with two simultaneous probabilities. If the dealer would place the roulette ball randomly on the wheel, the unfavorable repeat bias would be less present.

Even if *within* the *normal probability rule*, the bias could burn the casino out. A roulette player could play the top 10–12 pairings of each roulette number. For example, every time number *0* (*the green*) was drawn, the player would play the following 10 roulette numbers: 1, 15, 25, 9, 3, 20, 8, 30, 32, 33 (see the cyber report). Those numbers followed roulette number *0* in the next spin a total of 35 times out of 61 hits.

That represents *57%* of all cases when the roulette number *0* hit. It comes down to playing only the top pairings of each spin and winning 50% of the time! Play 12 numbers—and hit once, miss once. The total of 24 units in cost yields a win of 36 units straight up. That would be 12 units net win in 2 spins or an average of 6 units in net profit in every spin! A profit of 600 units per 100-spin sessions.

The problem is we don't know *in advance* which numbers will be the top 12 pairings for each roulette number! The skips of the pairings themselves must be determined as well. *Only a computer program can determine the skips and pairings in real-time—mission impossible inside a casino!*

Otherwise, of course, the roulette wheel could be extremely biased due to wear. Say, we analyzed the previous 1,000 roulette spins (recorded by hand), and we are certain of the bias. Meanwhile the casino executives have no clue of such bias! In my opinion, that is *not* a realistic expectation. The casinos always analyze the roulette spins and look for possible biases due to wear. I am not sure they perform pairing analysis because I have not given anybody my specialized roulette software. Still, there are several statistical methods to hunt wheel biases. The casinos use extensively the *chi-square test*. Question is what parameters and how many parameters do they analyze?

The analysis in this section was, for the most part, due to the sake of probability theory rather than gambling.

4. The Super Roulette Strategy

I was shocked to learn in 2005 that a licensed roulette strategy I sold in 2000 became freely available as the music at Napster in the past. At least one pirate collected over 300 roulette systems and made them available to everybody! I discovered three of my roulette systems in the collection. The first one, *Handy*, was formerly named *Handy-Brandy*. It is now known as the *Free Roulette System #1*.

The *Licensed* roulette system and *Limited* are one and the same. It is that old roulette strategy I sold in 2000 to a limited number of people. Many of them did not deserve my generosity. I suspect most of them were casino moles and/or disgruntled gambling authors who sell fraudulent roulette systems.

I decided to make my formerly *licensed roulette strategy* available as freeware. I viewed it as an act of justice. The pirates were already in the possession of my gambling systems illegally. Let honest people also get to know and perhaps use my ***Super Roulette Strategy***. My act was meant to be an equalizer. The original material follows.

The roulette systems in this collection are based on the Fundamental Formula of Gambling (FFG). The FFG is presented in detail on my Web page *Gambling Formula: Degree of Certainty, Probability, Mathematics, Chance*. You can download the page and print it. Although you do not need to memorize it, FFG can help you a great deal in understanding these systems better. The *Formula*, if understood well, can make clearer that probability events (or random events or gambling) actually follow precise rules.

A. The probability, *p*, for a *single-zero roulette* game is *p = 1/37 = .027.*

We apply the Fundamental Formula of Gambling to one value of *DC* (degree of certainty): *66.6%* (or *2/3* or *2 out of 3 cases*). For DC = 66.6%, N = 40.4 (rounded up to *41* spins). There is a 66.6% (2 in 3) chance that each roulette number will repeat after 41 spins. Equivalently, there is a 66.6% chance that the next spin will be a number that also appeared within the last 41 spins.

B. The probability, *p*, for a *double-zero roulette* game is *p = 1/38 = 0.026.*

We apply the Fundamental Formula of Gambling to one value of *DC* (degree of certainty): *66.6%* (or *2/3* or *2 out of 3 cases*). For DC = 66.6%, N = 41.5 (rounded up to *42* spins or 41 for single-zero wheels). There is a 66.6% (2 in 3) chance that each roulette number will repeat after 42 spins. Equivalently, there is a 66.6% chance that the next spin will be a number that also appeared within the last 42 spins.

The accompanying roulette software program *SuperRoulette.EXE* simulates spins for both types of roulette wheels. We'll use now only the final part of the report. After a very large number of spin simulations, the 42-spin value confirmed FFG. Also importantly, even smaller number of roulette spins confirm FFG for a DC = 66.6%. Let's use the Fundamental Formula of Gambling with confidence. That's what formulas are for.

We'll use this number (*42*) in correlation with limited-step martingales (bets based on a martingale progression). Basically, martingale represents a gambling betting method. It doubles up the previous bet until the last step of the betting. For example, a 4-step martingale using a 1-unit bet follows this path: 1 unit, 2 units, 4 units, 8 units (step 4, end of betting). Don't worry about using a martingale: the Formula makes martingale a closed betting gambling system (a system with a limited number of steps).

What you need is a notebook and a pencil. Write down the last roulette spins from the oldest one available to the most recent spin. Do not start playing until you have at least 42 spins on your piece of paper. I prefer a small notebook with 20 rule lines. Multiples of 10 or 20 make it easy to count quickly the number of roulette spins. Use the roulette report that follows as the template (rows and columns). The report preserves the fixed font in my command prompt software.

```
* 0-Roulette Winning Pattern *
W+ = Win; L-= Loss; x = 0 hit
Casino Spins File: HAMB0106.WH1
Number of REAL Spins Analyzed: 7990

Spin Number Hit Last Red/ Odd/ Low/ Last 1-12& 1-12& 13-24&
# Drawn After 25 Black Even High 41 13-24 25-36 25-36

Average: 18
StndDev: 11

1 11-26 sp. L-B o l W+ W+ W+ L-
2 18-20 sp. W+ r E l W+ W+ L-W+
3 32+ 18 sp. W+ r E H W+ L-W+ W+
4 22+ 43 sp. L-B E H L-W+ L-W+
5 6-9 sp. W+ B E l W+ W+ W+ L-
6 31+ 10 sp. W+ B o H W+ L-W+ W+
7 4-21 sp. W+ B E l W+ W+ W+ L-
8 30+ 9 sp. W+ r E H W+ L-W+ W+
9 27+ 48 sp. L-r o H L-L-W+ W+
10 21+ 64 sp. L-r o H L-W+ L-W+
11 14-23 sp. W+ r E l W+ W+ L-W+
12 16+ 6 sp. W+ r E l W+ W+ L-W+
13 2-2 sp. W+ B E l W+ W+ W+ L-
14 6+ 12 sp. W+ B E l W+ W+ W+ L-
15 2-34 sp. L-B E l W+ W+ W+ L-
16 31+ 223 sp. L-B o H L-L-W+ W+
17 30+ 60 sp. L-r E H L-L-W+ W+
18 16-48 sp. L-r E l L-W+ L-W+
19 33+ 49 sp. L-B o H L-L-W+ W+
20 26-146 sp. L-B E H L-L-W+ W+
. . . .
```

You can also download from my site two files with actual roulette spins recorded at the casino at Hamburg, Germany (Spielbank). The file HAMB00.DAT covers the period February 1–6, 2000. The other file, HAMB0106.WH1, covers the entire month of January 2006 (7,990 real-life roulette spins. I can only assume they are real or correct or accurate; it's impossible to verify such data). The data files are ASCII or simple text files. They look very much like the random output files generated by *SuperRoulette.EXE* in random mode (Random0.DAT or Random00.DAT).

The files contain roulette numbers (spins) only, one number per line. The casino *actuals* have also other data such as frequency and various statistics. The files also have dashed lines (—), which probably represent casino dealer changes.

HAMB0106.WH1 I edited and saved represents the best recording format: table by table. Do not mix the spins from different roulette wheels in the same file. Also, the most recent roulette spin always goes to the top of file while the oldest spin (casino result) goes to the bottom (bottom line). Each file line contains one roulette spin and ends with a carriage return (pressing Enter). Make sure you do not leave any blank lines: they are treated as roulette number 0 (zero). If your file is for double-zero roulette tables, be sure to type *37* instead of *00*.

The column *Last 26* reflects a DC (degree of certainty) of *50%* (it is *25* for single-zero casino roulette wheels). In half the situations, the next roulette number would have also appeared in the previous *26* spins. I decided not to use a betting system based on a DC = 50%. The losing streak can reach sometimes 10–13 spins in a row (in thousands of spins). Also, the losing/winning streaks are far more irregular. Instead, I chose to base my gambling betting systems on *DC = 66%* since the winning/losing streaks are a whole lot more manageable.

• ***System 1***: Applies to Report Column *Last 42 Roulette Spins*. The degree of certainty is *66.6%* that the next spin will be a number that also appeared within the last *42* spins. In most cases, there are 25–26 numbers to play since some numbers are repeats. The advantage of this system: the losing streaks (L or -) are far shorter, more regular, and less frequent.

From the opposite viewpoint, the winning streaks (W or +) are far longer, more regular, and more frequent. We can play a more efficient and safer martingale.

First, we need to keep a good record of the spins. We (you, if you will) need to make the record accurate and easy to read. *After* the spin no. 42, write down + or *W* if the number was a repeat from the last 42 spins or – (the minus sign) or *L* if the roulette number was not a repeat from the last 42 spins. Also, keep writing the numbers drawn. You'll write two rows such as what follows:

16,11,16,15,20,36,28,32,19,28,13,24,2,16,23,31,10,19,27, 8,26,27,00,19,16,12,36,18,9,8,30,6,14,17,25,12,8,3,17,18, 10,13,|7,14,2,19,7,24,32,2,25,11,13,29,17,25,25,14,26 . . . (first row)

—keep a few empty lines for more spins—

-++++++++-+-+++++ (second row)

The bar (|) after no. 13 marks the start of the betting for *System 2*. The first number drawn after you marked the start of the betting is 7. It does not appear among the previous 42 spins, so you write a – in the streak column. I prefer +/– rather than W/L because I can visualize easier the winning/losing streaks. The next operation is to write another bar after 16. It will make it easier to figure out the current 42-number roulette block. In this example, it will be between 11 (the number following 16) and 7.

The next number drawn was 14, which appeared also in the previous 42 spins. Write a + in the streak column. Write another bar after 11. The new 42-number block will span between 11 and the 14 just drawn.

So you always write down the very last number drawn and move the marking bar one number to the right. This way, you don't have to count the previous 42 spins. Place your chips on the numbers between the two bars marking the latest 42-number block. Of course, some numbers are repeats, so you only need to play each number once. In most cases, you'll only play 26 unique roulette numbers (sometimes 24, sometimes 27).

1. Roulette Systems Based on p=2/3 (66.6%)

S1.1. Applies to Report Column: Last 42. It is very rare that there are more than 2 consecutive groups of LL or longer than 2L. Therefore, at the end of the second LL or longer streak, we'll bet 1 unit on all the numbers drawn in the last 42 spins (24–27 numbers to cover).

In most cases, we'll win right away. If not, we wait for the next LL or longer streak. We'll use a power-3 martingale: 1-3-9-27, etc. When the LL streak ends, we'll bet 3 units on all the numbers drawn in the last 42 spins. It may be even less frequent to encounter another LL or longer streak. If it occurs, we may regret if we didn't continue the martingale with a 9-unit bet!

S1.2. Applies to Report Column: Last 42.
In reverse, we can bet on W groups longer or equal to 2 (WW, WWW, WWWW, etc.) Now we wait for 2 consecutive groups of single W. At the next W (after an L), we bet that it'll be another W. In most cases, we'll win right away. If not, we'll bet 3 units at another W after an L.

The safer variations on the last 2 roulette betting methods:

We can wait for 3 single W groups and bet directly $10 or $20 at the very next W after an L. There is no such occurrence in the report above. It might occur, albeit rarely. We can wait for three LL groups (2L or longer) and bet directly $10 or $20 at the very next L after a W. There is no such occurrence in the report above.

<u>Variations</u>
You can also bet directly $10 (or 2 units) after the second L in an LL roulette sequence. Or we can bet directly $20 (or 4 units) after the third L in an LLL sequence.

What would the cost amount to? It would be $10 x 26 numbers = $260.

The winning: $36 x 10 = $360
Profit: $100

Based on the Fundamental Formula of Gambling, these are the streaks for a probability, *p*, of .666. The degree of certainty, *DC*, is 95% that the losing streaks will be no longer than 3 (LLL). The degree of certainty, *DC*, is 99% that the losing streak will not be longer than 4 consecutive losses (LLLL).

The chance of coming across 3 consecutive LLL groups is 5% (0.05) to the power of 3 = 0.000125 (1.25 in 1,000).

The chance of coming across 3 consecutive LLLL groups is 1% (0.01) to the power of 3 = 0.000001 (0.001 in 1,000).

The 66.6% probability offers a tremendous advantage over probabilities around 50%. First and foremost, the losing streaks are definitely shorter. There is a 99.9% degree of certainty that the losing streak will be no longer than 6 when p=66.6%. That can make possible to use an intelligent martingale after a second or (more safely) third loss in a row (LL or LLL). In 999 out of 1,000 roulette spins, the L streak will go no longer than 6. Therefore, a martingale will go no longer than 4 or 5 steps. That is perfectly manageable within the minimum/maximum ratios at the roulette tables.

Secondly, the winning (W) streaks are definitely longer than the losing (L) streaks. It is very rare that two groups of single W will not be followed by W streaks longer than two (WW, WWW, etc.). We can reverse the martingale technique presented above. After the second single W (such as LWLLWL) we'll start another 4—or 5-step martingale.

We should also expect (high expectation, indeed) that W streaks of 4 or longer are common. Therefore, if we do not see such W streaks in the last 3–4 W streaks, we should bet they will occur soon. If there is a W/L string such as WWLWLLWLWWLLW, I can bet there will be another W and another one.

The disadvantage of the roulette systems based on p=2/3: the use of a power-3 martingale: 1-3-9-27-81.

The roulette systems based on $p = 1/2$ are using a power-2 martingale: 1-2-4-8-16.

• Systems 2, 3, 4: Apply to Report Columns 1–12 & 13–24, 1–12 & 25–36, 13–24 & 25–36 A B C

These roulette systems are a variation of ***System 1*** based on the 66.6% degree of certainty. At the same time, they are a reply to the fictitious roulette "system" presented in the James Bond novels and films. No doubt, the author of the James Bond novels, Ian Fleming, is a person with a strong interest in roulette, casino gambling in general. He noticed what I presented above: the winning streaks tend to be longer than 1 (WW or longer) for degrees of certainty above 50%.

Ian Fleming, however, has no clue on the mathematical foundations of such occurrences. James Bond plays *1–12 and 13–24*, betting large amounts of money. Since it is fiction, James Bond always wins at the roulette table! Now if any person in this world will walk to a roulette table and place bets on *1–12 and 13–24*, he/she will win a 2-to-1 payoff 65% of the time at French roulette (64% playing American roulette).

In other words, if you would play this way at any roulette table 3 times in a row, you will win a 2-to-1 payoff 2 times (not necessarily twice in a row). You will also lose once, roughly. That's a far cry from the 100% winning insinuated by Fleming in his book *Casino Royale*.

You should never ever play in that manner at the roulette table! You should enter the game at favorable moments, as presented in ***System 1***. When such favorable moments occur, get ready to use a limited-step martingale (no longer than 4 or 5 steps). The betting sequences are 1-3-9-27, etc. This is determined by the 2-to-1 payoff when playing 12 roulette number groups (*douzaine*).

When using this betting plan, you are only allowed to place bets on numbers 1 to 36. You are not allowed to include 0 (or 0 and 00) in the groups of numbers you bet on. The 36 numbers can be divided into three groups: A (1–12 and 13–24), B (1–12 and 25–36), C (13–24 and 25–36). Each of the three groups (A, B, or C) has an equal probability: *p = 64%* (in the American roulette) or *p = 65%* (in the French roulette). The two probabilities are very close to the degrees of certainty we used as the foundation of SYSTEM 1 (66.6%). Consequently, we will encounter virtually the same composition of the W and L streaks.

Systems 1.3, 1.4, 1.5 do not require a long tracking session at the roulette table. The table will show the last 20 spins, so you can start betting immediately. You can choose to play only one of the three *double-douzaine* groups: **A**, **B**, or **C**. Or you can play two of the groups or all three groups. It is best, however, to keep track of all three groups. Only one of the three *double-douzaine* groups will have a stronger advantage over shorter tracking sessions.

It is easy to memorize what the groups represent:
A = first douzaine + second douzaine (1–12 & 13–24)
B = first douzaine + third douzaine (1–12 & 25–36)
C = second douzaine + third douzaine (13–24 & 25–36)

You will write the roulette numbers drawn in the first row and the W+/L—streaks in the next 3 rows:

```
11 21 7 32 25 2 2 29 23 7 29 28 7 34 27
A: + + + - - + + - + + - - + - -
B: + - + + + + + + - + + + + + +
C: - + - + + - - + + - + + - + +
```

We apply the same betting principles as presented in ***System 1***. The structures of the W+/L- streaks are very close to the W+/L- strings encountered in ***System 1***. Keep in mind this mathematical rule. When the probability, *p*, is significantly above 50% (10%+ above 1/2), most of the winning streaks (W+) are 2 in length or longer (WW, WWW, WWWW, etc). On the other hand, most of the losing streaks (L-) are singles:

LWWLWLWWWLLWLW . . . When betting, we know that only rarely will we encounter the following:
~ more than three consecutive groups of LL or longer (LLL, LLLL . . .). Using the +/—notation and a real report generated by SuperRoulette.EXE, a W/L streak looks like this: ++-+++--+--+++--++-+

You can see three consecutive groups of LL (—). The very next L group was single (-), ending the L streak longer than or equal to 2.

~ more than three consecutive groups of single W+. Using the +/—notation and a real report generated by *SuperRoulette. EXE* roulette software, a W/L streak looks like this: --+++++----+-+---+++

You can see two consecutive groups of single W+. The very next W group was longer than 2, ending the single W+ streak.

Based on the two probability facts, we can devise two betting methods. They are similar to the methods described in SYSTEM 1.

• ***S234.1.*** We check to see which of the groups A, B, C has two or more consecutive groups of LL or longer. At the end of two consecutive groups of LLW or longer (LLLW, LLLLW, etc.), we place one unit bet on a douzaine and another unit bet on the other douzaine of the A or B or C group. Let's look at a practical example. Group A has had the following streak: ++-+++--+--+++--++-+. Or using the W/L notation, WWLWWWLLWLLWWWLLWWLW.

You can notice three consecutive LLW groups. The second group is LLWWW. Then an L (or -) follows. You start betting right there, after the first L. You know that there is a high probability to get a single L. In this example (a real case though), you lost your roulette bet. The first L was followed by another one. The L streak ended with WW. Another L follows. You start betting again after that first L. The probability is even higher now for L to be single (that is, it will be followed by W). Bet now 3 units on the first roulette douzaine of the group and 3 units on the other douzaine of the group. In this case, you won a 2-to-1 bet. You won 3 x 3 units = 9 units. Your cost was 3 units + 3 units = 6 units. The net gain is 3 units. But you lost the previous bet: 1 unit + 1 unit = 2 units. Deducting 2 units from the gain above, your final gain is 1 unit.

• ***S234.2.*** We check to see which of the groups—A, B, C—has two or more consecutive streaks of single W. At the end of two consecutive groups of WL, we place one unit bet on a douzaine and another unit bet on the other douzaine of the A or B or C group.

Let's look at a practical example. Group A has had the following streak: —+++++—+-+—+++. Or using the W/L notation, LLWWWWWLLLLWLWLLLWWW. You can notice two consecutive LWL groups (isolated W or +). The second group is LWLLL. Then a W (or +) follows. You start betting right there, after the first W. You know that there is a high probability to get a W longer than one (WW, WWW, etc.).

In this example, you bet $10 + $10 (after that first W) that another W will follow. It actually did happen, so your payoff was $30 ($10 profit). If it did not happen (as in case S234.1 above), you would continue with a double-up base-3 bet: $30 + $30 = $60. In most cases by far, you will win a $90 payoff. The profit would be $90 – $60 – $20 = $10.

Do not fear reality. If you lost again, you would continue with a double-up base-3 bet: $90 + $90 = $180. In most cases, you will win a $270 roulette payoff. The profit would be $270 – $180 – $60 – $20 = $10.

You can still lose another step and double up within the roulette table maximum bet. Losing that many steps in this manner is so extremely rare that it may take to play the roulette for many, many thousands of spins in a row.

The safer variations on the last two betting methods:

We can wait for three single W groups and bet directly $10 or $20 at the very next W after an L.
We can wait for three LL groups (2L or longer) and bet directly $10 or $20 at the very next L after a W.

Variations
You can also bet directly $10 after the third L in an LLL sequence.
Or we can bet directly $20 after the fourth L in an LLLL sequence.

2. Roulette Systems Based on p=1/2 (50%)

We apply the Fundamental Formula of Gambling to a value of DC (degree of certainty): *50%* (or 1/2).

For DC = 50% => 25 (rounded up to *26* spins). There is a 50% (1 in 2) chance that each roulette number will repeat after 26 spins. Equivalently, there is a 50% chance that the next spin will be a number that also appeared within the last 26 roulette spins.

I was reluctant to recommend roulette systems based on *p=1/2*. People may be tempted to use martingales that a losing streak will end within 4 or 5 steps. Please do not do that. A losing streak can reach 13–15 spins (in thousands of spins)! Instead, I will show you a more intelligent type of betting. They are similar to the systems in category 1. They also use a power-2 martingale (1-2-4-8) since the payoff is 1 to 1.

The report columns for this betting category are as follows:

Last Red/Odd/Low/
26 Black Even High

S2.1. It is very rare that there are more than three consecutive groups of LL or longer than 2L. Therefore, at the end of the third LL or longer streak, we'll bet 1 unit on all the roulette numbers drawn in the last 26 spins (17–20 numbers to cover, 18 in most cases). In most cases, we'll win right away. If not, we wait for the next LL or longer streak. We'll use a power-2 martingale: 1-2-4-8, etc. When the LL streak ends, we'll bet 2 units on all the roulette numbers drawn in the last 26 spins. It may be even less frequent to encounter another LL or longer streak. If it occurs, we may regret if we didn't continue the martingale with a 4-unit bet!

S2.2. In reverse, we can bet on W groups longer or equal to 2 (WW, WWW, WWWW, etc.). Now, we wait for 3 consecutive groups of single W. At the next W (after an L), we bet 2 that it'll be another W. In most cases, we'll win right away. If not, we'll bet 4 units at another W after an L.

The safer variations on the last two betting methods:

We can wait for 4 single W groups and bet directly 4 units at the very next Win after a Loss. We can wait for four LL groups (2L or longer) and bet directly 4 units at the very next L after a W.

Systems 2.3, 2.4, 2.5 cover the even-money bets: **Red/Black, Odd/Even, High/Low**. They follow the same rules as S2.1 and S2.2. In this case, however, we do not track W/L, but B/r, o/E, H/l.

• Let's take the Black/Red bet. Write first the numbers drawn:

36 10 29 25 16 1 22 4 7 23 26 4 10 29 21 27 16 15 6 37 19 22 29 36 18 16 30 37 35 10 37 15 16

Next, write down what the roulette number represents: *B* for black, *r* for red, and *x* for 0/00:

rBBrrrBBrrBBBBrrrBBxrBBrrrrxBBxBr

Bet 2 units after the first B in the fourth B group 2 or longer (it is BB after the third r in the row above). We lost because the BB streak did not end there. The next B (black at roulette) group of 2 Bs or longer meant we lost again; this time we lost 4 units. Another BB group follows, and we bet 8 after the 1st B in the group. There is an x, then another B. We win this time. A safer variation is to wait for four consecutive BB or longer groups. They occur far more rarely however.

We can do the same thing betting on r. You can notice there are 3 rr or longer groups (rrr, rr, rrr). They are followed by a single r (BxrB).

• As in ***System S1.2***, we bet on the continuation of singles (single B, single r, etc.). Let's look at this Odd/Even streak:

oooEoExoEoEEEExoExoEoEEoEooE.

We bet 2 units after the third single o. We lose. More single o groups followed, and we used the following martingale:

4 (loss) – 16 (loss) – 32 (loss) – 64 (WIN). Then 64 x $5 = $320, an acceptable maximum limit at most roulette tables. The roulette game in Garmisch-Partenkirchen, Germany, has a maximum limit of DM 6000 for even-money bets (for a DM 5 minimum limit).

Can you keep track of all the roulette betting systems presented here? It is difficult in the beginning, but not impossible to write down the roulette numbers in rows and columns. Practice in a real casino, at a real roulette table, helps a great deal. First practice then play roulette with real money. The advantage of keeping track of all betting systems: CHOICES. You have plenty of options to choose from in a short time. You don't need to track a long number of spins until one of the favorable situations will come up. It is very likely that you will find yourself in a betting situation right after writing down the first fifteen numbers displayed at the roulette table (the marquee). A team of two players at the same roulette table is even more efficient. I tried it, and the efficiency was clearly better.

I will show you again a notebook fragment from my real play in Atlantic City, July 22, 2003. I limited my tracking to fewer columns as I realized I was not fast enough to cover all bases.

Spin	In/De	StD	D12	D13	D23	K12	K13	K23
21			+	-*	+	-	+	+
4	-	-	+	+**	-	+	+	-
17	+	-	+*	-	+	+	-	+
4	-	-	+	+	-	+	+	-
15	+	+	+	-	+	-	+	+*
22	+	+	+	-	+*	+	+	-
23	+	+	+	-*	+	+	-	+
20	-	+	+	-*	+	+	-	+
8	-	-	+	+**	-	+	-	+
33	+	-	-	+	+	-	+*	+

Here is the interpretation of this, really, winning table.

The **Spin** column represents the actual roulette number. **In/De** represents an *increase* or a *decrease* relative to the preceding spin. For example, the first number recorded was *21* and the next roulette number was *4*; thus, a *decrease* and the – (minus) sign.

D12	D13	D23	K12	K13	K23

These represent the double dozens and double columns as explained earlier in this section.

StD represents the good ole standard deviation. You can see the *statistical standard deviation* for the Hamburg data file in the previous report: *11*. The real results file has about 8,000 roulette spins—a significantly high amount of data for roulette. However, I calculate the *FFG deviation* as I remarked in the chapter dedicated to standard deviation. *The **FFG deviation** represents the **number of trials,** **N,** for a **degree of certainty equal to 25%** for any **probability,*** **p**. We calculate ***p*** at roulette as ***1/37*** (single zero) or ***1/38*** (double zero). The ***FFG deviation*** is the same: ***11***.

I use the *FFG deviation* as a betting parameter. This parameter can be applied only in conjunction with the **In/De** parameter. I must determine first if the next roulette number will be an *increase* or a *decrease*. Let's look at the real number **15**, which was a winning situation for me. I expected an *increase*, and I was right. I also expected the *increase* to be *within FFG deviation*.

I had failed in the previous spin: **4** represented a decrease from **17**, but the difference was 13. Therefore, it was outside the standard deviation of 11. This time, number **15** was an increase of 11 from the previous spin (number **4**). I had lost

11 units in the previous spin. I paid 11 more for this spin, but this time I won 36: a profit of 14.

I applied this **StD** method four times in this particular span of 10 spins. I won twice and failed twice. I should have increased my bet after a failure, but I didn't. The method still gained a $140 profit (14 x 2 x $5).

The * sign marks the betting situations on the double dozens or columns. The columns are harder to track as they don't consist of contiguous numbers. Hence, my betting on the Ks was lighter. Also, my focus was on the **StD** parameter. I started casually with the $5 minimum bet in the **D13** double dozen. I lost. I triple-martingaled the next spin and won (the ** means a martingale step). Next, I bet on a long streak of **D12**. It was long, indeed, but I bet only once on it. As earlier in this section, the p=2/3 parameters do encounter long winning streaks at times. It happens to at least one of such parameters within the same range of analysis (roulette spins).

Later in the game, I would bet $25, $75, $225. The lady dealer became exasperated with my play. She almost sounded like crying. *"If you play like that, you'll be here forever and never lose!"* Her boss tempered her somehow. He also placed two shills at my table. The two young ladies did their very best to *unfocus* my attention from my notebook to their faces and designer T-shirts! I left the table soon. I was happy. Yet the rumors made the rounds of the Atlantic City casinos. Later in the evening, at Taj Mahal, a former football lineman already knew that I had chipped in $2,000 at Trump Plaza and won quite a bit! He really kept me away from the roulette table. He said he had a job to do as a bodyguard! His man was playing big at that table (an Asian American). The story is there at my Web site.

5. Roulette Software: Significant Titles

I list here the most significant roulette software titles that you can easily obtain from my Web site.

• *SuperRoulette.exe* is possibly the most advanced and most promising act in roulette software programming. The software is the indispensable companion to the incredible *Super Roulette Strategy with the Best Roulette Systems Ever Released*—as presented at SALIU.COM and in this chapter.

• *RouletteHemis.EXE* is a superior roulette spins analyzer regarding the slot (half) location and the *Birthday Paradox*. The roulette wheel is divided into two semicircles (hemispheres or halves). The *inside* sector consists of the number drawn plus 9 slots to the left and 9 slots to the right. The opposite half of the roulette wheel represents the *outside* hemisphere.

• *SPINS.EXE* is an excellent roulette spin generating application, plus a statistical analyzer of the roulette spins. This program is also a great complement to the *Free roulette system #1* presented in this chapter. According to the system, it is recommended to play after 2 consecutive "-" in the "Result +/-" column.

• *Roulette.EXE* does what SPINS.EXE does, and then some. This program adds a new system to the *Free roulette system #1* presented here. Instead of tracking only the last 5 spins, Roulette.exe tracks also the last 15 spins as displayed on the roulette marquee (an application of the *Birthday paradox*).

• *SkipSystem.exe*: Software to automatically create roulette, horse racing, lottery, gambling systems derived from skips and based on the FFG median.

• *BellCurveGenerator.exe* generates combinations within the FFG median bell. The Fundamental Formula of Gambling calculates the median automatically. The application handles just about any game: pick 3, pick 4, lotto-5, lotto-6, lotto-7, Powerball, Mega Millions, EuroMillions, horse racing, roulette, and sports betting (including the famous European soccer pools and American sports teams).

The download location for the roulette (and other games) software category is *saliu.com/free-casino-software.html*

Chapter XIII

AN INTRODUCTION TO LOTTERY MATHEMATICS

1. The Protomathematics of Lottery

Probably, *protomathematics of lottery* starts with K. J. Nurmela and P. R. J. Östergård at the Helsinki University of Technology. Their publication is titled *Constructing Covering Designs by Simulated Annealing.* You can read this PDF document
http://citeseerx.ist.psu.edu/viewdoc/download?doi=10.1.1.48.3798&rep=rep1&type=pdf.

The first attempt was followed by Pak Ching Li and G. H. John Van Rees at the University of Manitoba, Canada. The main scientific paper is titled *"New Constructions of Lotto Designs."* You can find and read two online PDF documents:
www.cs.umanitoba.ca/~vanrees/upbd.pdf
www.cs.umanitoba.ca/~vanrees/fau4.pdf

They all deal with *lotto designs* and techniques for constructing *lotto designs* and determining *upper bounds* for *L(n, k, p, t).* *Lotto design* sounds more academic than the layperson's term!

The *academia* people also use another fancy term: *covering designs*. The *lotto designs* are better known by laypersons as *lotto wheels* or *abbreviated (reduced) lotto systems*.

A *lotto wheel* is simply a set of a *reduced number of lotto combinations* that assure a certain *minimum guarantee* at the cost of forsaking the highest prize.

The would-be lottery mathematicians attempt to discover formulas for constructing the tightest possible lotto wheels. The professors come up with all kinds of approximations that look like mathematical formulas (by highly symbolized notation and heavy dosage of jargon). The margin of error, however, is as large as the area of Canada!

You might want to know the *original* mathematics of *approximation*, especially as applied in theory of probability. You might have heard of *Chebyshev's inequality*. I give a most simplified, but most comprehensible, interpretation of the famous inequality. For example, the approximation inequality guarantees that the probability to get *heads* in 10 tosses is higher than 75%. The assessment is correct, but the margin of approximation is gigantic. Au contraire, the Fundamental Formula of Gambling (FFG) restricts the approximation to the *minimum minimorum*. As in the example above, the degree of certainty of 75% to get *heads* requires just 2 tosses. For a number of tosses equal to 10, the degree of certainty reaches 99.9%.

The professors analyze one lotto wheel, a famous one by now: *49 lotto, 6 numbers with the* ***3 of 6*** *minimum guarantee*. In cuckoo speak, *LD (49, 6, 6, 3)=163*. That is, the guarantee of the lotto wheel for *6-from-49* numbers is satisfied in *163* lines (or 163 6-number combinations). The flaw is evident right from the start. It is not one group of 49 numbers, but *two separate* groups of numbers: *22* and *27*, respectively. None of the numbers in the first group meets with any number in the second group.

The task of analyzing a flawed lotto wheel raised to the status of *lottery mathematics* sounds funny to me. But hey, we got to start somewhere! Many things start like that! The lottery professors even invoke a true mathematician, Euler, in their research. They sow the pages with a plethora of theorems and formulas.

We have also met in this book a special probability formula: *hypergeometric distribution*. The hypergeometric probability distribution formula calculates the probabilities of getting ***k** of **m** in **p** from **n** numbers*. In layperson's words, the probability or odds to hit 3 (k) of 6 (m) winners, when I play a pool of 18 (p) numbers in a lotto game with a total of 49 (n) numbers in the field. The answer in this particular example is *1 in 3.8* (or approximately once every 4 lotto drawings). But if we only play a single combination of 6 numbers instead of an 18-number pool, the *3-in-6* odds are *1/56.66* (or approximately once every 57 lottery draws). The calculations are easily available to everybody by running the online *random generator, odds calculator* from my truly mathematical site!

The wrong assumption of the lottery professors is the possibility of constructing a lotto wheel (lotto design sounds better?) consisting of 57 lines (6-number combinations) to satisfy the 3-in-6 minimal guarantee. That will never happen! No matter how many highly symbolized formulas and theorems a university paper would print! The number *57* here is not probability, is not degree of certainty. It is the number of trials! These are the three fundamental elements of random phenomena, including lottery and lotto games. The three elements are embodied in the Fundamental Formula of Gambling (FFG): Probability (p), degree of certainty (DC), and number of trials (N).

We don't need a *lotto design* to hit the lottery according to the hypergeometric distribution formula. Just play 57 unique 6/49 lotto combinations. Choose them randomly (as by running my online generator) but be sure to strip of any duplicates. You'll notice that, for example, the 57-line set of random lotto combinations does not have a *3-in-6* winner in a particular lottery drawing. But other times, you will get two 3-in-6 winners per draw. If you play the 57-line set in 10 lottery draws, you will record 10 3-in-6 winners. So what's the purpose of lotto designs? The higher costs such as spending time, effort, extra money for playing, etc.?

No lotto design will come even close to the mathematical odds, as far as the number of combinations is concerned. I found one exception: the 10-number case for *4 of 6 [LD(10, 6, 6, 4)]*, in 3 combinations. That is nothing but a simple exception; it is not mathematics. I wrote the only software that can generate reduced lotto systems that do *not repeat* a certain amount of numbers from previous combinations. That is, all the numbers in the reduced set (or covering design or lottery design) do not have **k** numbers in common. In the *3-in-6* case, no combination in the set shares 3 or more numbers with any combination in the set. Before I released my software, the best *LD(10, 6, 6, 4)=5* consisted of 5 6-number lines. The best *LD(11, 6, 6, 4)=11* consisted of 11 6-number combinations; my software does it in 5 lines. But only *LD(10, 6, 6, 4)=3* comes closest to the mathematical odds (probability to get *4 in 6*). For any other *lotto design* or *covering design*, the amount of lines in the reduced sets is ***randomly*** greater than the mathematical odds.

The aforementioned lottery professors abundantly use the < (less than) and > (greater than) mathematical operators. When things are not sure, those operators come to the rescue! Let's say the minimal size of *LD(49, 6, 6, 3)* must be > (greater than) 57 combinations! Of course, but it can't be proven by a mathematical formula! How about *LD(18, 6, 6, 4)*? It has to be greater than 19. Of course, but it can't be proven by a mathematical theorem!

The true question is by how many lines *greater than exactly*? They don't know because the number of lines is *randomly greater* than the calculations. Thus, your best bet is to play a set of random combinations equal in size to the odds calculated by the hypergeometric distribution formula for that particular lotto prize. And you'll be better off financially. I demonstrated that truth with real data in UK National Lottery by comparing the 163-line lotto wheel to a set of random lotto combinations. Read the Web pages dealing with lotto wheels at my Web site, especially the test of real data, Saliu.com/bbs/messages/11.html.

Here is how the *random set* fared:
Total Hits: 0 "*6 of 6*"; 0 "5 *of 6*"; *41* "4 *in 6*"; ***666* "3 in 6"**
Total Cost: 642 x 57 = *36,594* units
Total Winnings: (666 x 10) + (41 x 100) = *10,760* units
Net Loss: *25,834* units

Here is how the *163-line wheel (design)* fared:
Total Hits: 0 "*6 of 6*"; *1* "*5 of 6*"; *104* "*4 in 6*"; ***1872* "*3 in 6*"**
Total Cost: 642 x 163 = *104,646* units
Total Winnings: (1872 x 10) + (104 x 100) + (1 * 3000) = *32,120* units
Net Loss: ***72,526*** units

You'll find there all the tools you need to test your own data: my free software.

The only precise action in this realm of mathematics is generating all possible combinations in a lotto game. The lotto combinations are a particular set of numbers (one of the four types of sets as seen at the beginning of this book). My freeware PermuteCombine.EXE generates every imaginable set of numbers (words or text as well), both in lexicographical order and in random manner. There is a total of 13,983,816 combinations in a 6-from-49 lotto game. My free software will generate exactly 13,983,816 combinations in lexicographic order (from 1, 2, 3, 4, 5, 6 to 44, 45, 46, 47, 48, 49). The lottery professors or lotto wheel aficionados will call those 13983816 combinations a 6-of-6 lotto design or a full 6/49 lotto wheel! The contradictions in terms are blunt. *Design* or *wheel* implies *reduction* (i.e., a reduced set of numbers that satisfy a condition).

As per above, I wrote also lottery software to generate a wide variety of lotto wheels or reduced (abbreviated) lotto systems. The program names read something like *WheelCheck.EXE*. You can run them for free if you register as a download member for a nominal and reasonable fee. That formidable lottery software is also designed to *verify* lotto wheels for *missing* combinations and generate reduced lotto systems.

The source code of that software is so much sought after! I got tired of how many requests I received. Lots of people even go to public forums where I contribute from time to time. The requests reach the pathetic form at times. The source code for that type of software can save the humanity from many disasters! I heard that the pharmaceuticals need that type of software to make sure that the same ingredients were not combined more than a minimal frequency, etc. I know for a fact that even members of academia resort to such type of requests. My case: I *must patent* my ideas and software before I reveal everything I know, including the algorithms in my software. One huge problem: ***money***. It would take more than a million dollars to patent all my knowledge—and that just by doing all the dirty work myself. In my case, it's impossible to patent everything I discovered or programmed without a team of lawyers. And who would pay for all the trouble? I possess millions of dollars only in my far-fetched dreams.

My lottery wheeling software treats all lotto numbers as fairly as possible. That feature is called *balance* (i.e., the numbers appear *fairly equitably* in the reduced system). I label the lotto wheels generated by my software as *balanced*. For the *1 of 6 in 6 from 49*, *WheelCheck6.EXE* generates 8 and always 8 combinations. Of course, the hypergeometric odds are *1 in 2.42*. That is, some players expect a 3-line lotto wheel! That is not possible. Just look at the 8-line set:

1 2 3 4 5 6
7 8 9 10 11 12
13 14 15 16 17 18
19 20 21 22 23 24
25 26 27 28 29 30
31 32 33 34 35 36
37 38 39 40 41 42
43 44 45 46 47 48

The number 49 is missing, but the design still covers 1 of 6 from 49. The 8-combination system covers up to 53 numbers. The incomplete combination 49, 50, 51, 52, 53 will need one more number to make it a 6-number lotto combination. It will be any of the 48 numbers in the 8-line "1 of 6" lotto wheel. The system is balanced in a proportion equal to 98%; 48 of the 49 numbers are equally distributed (one time each). One number (no. 49) is missing. This lotto system cannot be compressed (reduced) any further. By eliminating just one line, we destroy the 1 of 6 from 49 guarantee.

Compression or reduction is possible for multiple-number guarantees: from 2 of 6 from 49 to 5 of 6 from 49. Compression or reduction is no longer possible when we reach the 6 of 6 from 49 condition. The 13,983,816-combination set cannot be reduced by one single line!

The "2 of 6 from 49" lotto design generated by the WheelCheck6 program consists of 48 lines. The numbers are fairly equitably distributed, but there is a bias toward the first 2 numbers. The bias is caused by the fact that the lexicographical lotto generation starts with no. 1. If we do the generation in descending number (starting at no. 49), then the bias would favor the numbers 49 and 48.

The "3 of 6 from 49" lotto design generated by the WheelCheck6 program consists of 514 lines. The numbers are fairly equitably distributed, but there is a bias toward the first 3 numbers. The bias is caused by the fact that the lexicographical lotto generation starts with no. 1. If we did the generation in descending number (starting at no. 49), then the bias would favor nos. 49, 48, and 47.

I was able to reduce this lotto system down to 416 combinations. The minimal guarantee was preserved while the balance was improved. The procedure is painstaking however. First, I generated all 13,983,816 combinations in lexicographical order. It is a huge file (it can be also downloaded from my site)! Next, I ran another great piece of software I created: *Shuffle.EXE*. Among other functions, the program *shuffles* a text file. That is, the lines of a file are distributed in a *random* manner. (The randomized 13,983,816 combinations can be also downloaded at my software site.)

The shuffling procedure is easier for 5-number lotto files, for example. After one shuffle, you run another great free program of mine: *WheelIn6.EXE*. That program creates wheels from text files containing lottery combinations. You can shuffle again and again and again until you realize that you can't shrink the lotto wheel any further.

Another method is to start the generating with the Wheel.EXE programs (as in the Pick.EXE and especially Bright.EXE integrated lottery software packages, F8 key). You stop the lotto wheeling programs when everything seems to have come to a halt. Use the lottery wheel output file and input it to WheelCheck.EXE. Combine the new output with the output file generated by Wheel.EXE. You will have a complete lotto design assuring the minimum guarantee and a good balance.

The lotto designs or wheels created by these somehow-complicated methods are called *balanced* and *randomized*. The shuffling procedure assures that no lotto numbers are biased in any way. The lotto wheels are balanced to higher degrees while maintaining the minimum guarantee or condition. The *balanced and randomized* lotto wheels do guarantee higher winning per comparable cost because of our good old friend, *standard deviation*.

However, there is no formula that can calculate the number of lines (size) of the lotto design. The design sizes differ catastrophically from set to set! Looks like all the professors in lottery designs did was to fight the windmills!

The only way to improve the designs (the size of lotto wheels) is by shuffling or randomizing. The process does not follow any mathematical formula or procedure. It is a trial-and-error chore. The only thing to express for sure is this: a lotto design will always have a larger number of elements than the mathematical calculations. That is not science; that is not mathematics. That is a simple observation caused by a painstaking trial-and-error chore.

2. Mathematics of Lottery

Perhaps a bold statement: The true *mathematics of lottery* starts here and now. We go back to the Fundamental Formula of Gambling (FFG) right away. That's how I started the lottery strategy back in 1997, with a *real mathematical formula*, not vague approximations. The *equal to* sign (=) reaches the truth significantly closer than < or >.

Step One: One Number at a Time

It was the very beginning of my Internet experience (Saliu.com/LottoWin.htm). Very importantly also, it signed the birth certificate of lottery mathematics. The main lottery strategy page started this way:

> *FFG has a column p=1/8 that describes exactly a lotto game drawing 6 winning numbers from a field of 48 numbers. "6 divided by 48" is 1/8 or 0.125. That's how you calculate the probability p, when considering one lotto number at a time!*
>
> *Evidently, each lotto or lottery combination has an equal probability* ***p*** *as the rest, but the combinations appear with different frequencies. The FFG median is the key factor in the biased appearance. The lotto numbers tend to repeat more often when their running skip is less than or equal to the "probability median". The "probability median" or* ***FFG median*** *can be calculated by the Fundamental Formula of Gambling (FFG) for the* ***degree of certainty DC = 50%****. This revolutionary premise constitutes the backbone of the lottery and lotto strategy that follows.*

As seen above, the "1 of 6 from 49" lotto design required 8 lines. There is more to it. *Ion Saliu's Paradox of N Trials* demonstrates that if we randomly generate 8 lotto 6/49 combinations, only 63% of the numbers will be unique; the rest, 37%, will be repeats. The FFG median for this case is 6. If we generate 6 random lottery combinations, only half (50%) of the numbers will be unique. Equivalently, half of the 48 (or 49) lotto numbers will have not come out in 6 drawings.

This lottery strategy, as it was the case with the "1 of 6 from 49" lotto design, doesn't offer anything in the way of further reduction. We know now that after 6 drawings, only 50% of the lotto numbers came out. We also know that after two more drawings, 63% of the lotto 6/48 (let's say 6/49) would have come out. Thus, the drawings nos. 7 and 8 will offer 13% more lotto numbers.

In absolute terms, 13% represents 6 or 7 lotto numbers (out of 48 or 49). There were 24 (or 25—the half) numbers that came out in the previous 6 draws. FFG expects 6 new numbers to come out in the next 2 draws. The 6 numbers can be distributed over the next drawings in patterns like 0-6 or 6-0 to 15 or 5-1, but more likely 3-3. If we play only the numbers from the last 6 lottery drawings, we expect to hit 3 lotto winners. But things could be drastically different at (rare) times. As my lottery strategy page demonstrated, real lottery results showed that all 6 winners came from the previous 6 drawings, even from the last 5 or even 4 lotto draws!

Let's always give credit when credit is due. *Gail Howard*, formerly a stockbroker, noticed and published before me the tendency of lotto numbers to repeat from the most recent drawings. However, I was applying the method while in Romania, long before I had even heard the *Gail Howard* name! Not to mention that my lottery strategy applied not only single lotto numbers, but also pairs of numbers. Gail Howard did not apply any mathematical method or computer programming. She had analyzed one low-odds lotto game (6 from 38 or so), but she generalized her findings to all lotto games. She relied only on observation. Her method was entirely manual (pencil-and-paper). After selecting 15–20 numbers, the lottery player would *manually* wheel the numbers. She sold a book that mostly consisted of lotto wheels *manually* developed by a Bulgarian mathematician in the 1960s, Dimitrov. Gail Howard headlined her book as being based on the *Dimitrov lotto wheels*. I will always say this 100% sincerely: give to the pioneers what is due to pioneers!

Since no much reduction is possible, this incipient lottery strategy is the weakest. It was the one that I told you about four score or so pages ago. I made two economists win substantial money in the 1980s Romania. It is the strategy I applied with my Puerto Rican farm mates when I played the lottery in the United States for the first time (1985). I abandoned that method by the end of the 1980s.

Step Two: Two Numbers at a Time (Pairs)

I developed a more powerful lottery strategy three or four years after the incipient lottery strategy. I called this second coming the *wonder grid*. The lottery wonder grid considered *pairs* of lotto numbers instead of single lotto numbers. There was plenty of statistical evidence that the lotto numbers showed strong biases in pairing with other lotto numbers. The top-5 pairs for each lotto number came out a lot more frequently than the rest of the pairs: around 50% of all pairing frequency. We witnessed the same phenomenon when we analyzed the game of roulette.

The weakness of the *lottery wonder grid* was nondiscrimination. It treated equally all lottery numbers. I discovered, at a later time, the science of *positive discrimination*. You can read much more right at my Web site (where else?). We found out at *Step One* that the lottery numbers do not come out with an equal frequency. As a matter of fact, a few lotto numbers come out 6 times in 50 drawings while other numbers do not show up! On the other hand, the *wonder grid* treated every lotto number equally. It played each and every number and its top pairs.

I discovered an important zone of the area of lottery pairings I named *least pairings*. That discovery preceded the *Incipient Lottery Strategy* described by *Step 1*. The *least pairing* was one of the earliest *filters* (or restrictions) in my lottery software (beginning 1988).

The term *least pairings* is relative. We can fully define it by establishing the value of *least*. Least refers to the minimum threshold of the pair frequency. The default value I established for least in my lottery software is *0* (zero). That is, every lottery pair with a frequency of *zero* (no show) is considered to belong to the *least pairings restriction* (filter). But the least pairings filter is validated sometimes by an upper limit higher than zero. There are drawings when the least lottery pair has a frequency higher than 3–4, even higher! Setting the lottery filter that high absolutely devastates the lotto odds! Even a least pairing equal to 0 can reduce millions of lotto combinations!

Let's do some common sense mathematics regarding Step 2 and the *Least Pairings in Lottery*. How many pairs or pairings are there in a lotto game, say 6 from 49? There are two easy methods to calculate. The 49 numbers can be paired in C(49, 2) = [(49 * 48) / (1 * 2)] = *1176*. The lotto game draws 6 numbers; therefore, total pairs in the draw is C(6, 2) = 15 pairs. In final analysis, the 6/49 lotto game yields 1176 / 15 = *78.4* or approximately ***79*** integer elements.

The other method does it in one step, very much like calculating the lotto combinations: *[(49/6) * (48/5)] = 78.4*.

Those are *not* real elements. They are *derived elements* to help us perform probability calculations. We did it easier at Step 1 when we calculated the probability for singular lotto numbers (e.g., 1/8 or 1 in 8). Probability is what we need first and foremost. The formula above can be reworked to calculate pair probabilities directly: [(6/49) * (5/48)] = 0.012755 or 1 in 78.4.

We shall run at this point the most superb probability and statistics software: *SuperFormula.EXE*. We need to calculate the *FFG median* for lottery pairs. We should choose "Option 2: The program calculates p." I suggest we use *10* for the first element of p and *784* for the second element (for more tightly accurate results). The result for FFG median (DC = 50%) is something like 54. So half of the lotto 6/49 pairs will come out in 54 drawings. Hold on! The *WheelCheck6.EXE* program generated a lotto wheel that covers 2 of 6 from 49 in 48 combinations! Actually, that wheel can be reduced to 30-something lotto combinations while preserving the minimum guarantee! Well, that's the power of reduction!

We can generate the lotto-pairing file by running another piece of great lottery software: *Util.EXE* (a generic name, it can be Util-6 for 6-number lotto games). The *Stats* function generates a plethora of frequency reports, including for pairs. The program automatically creates a distinct file dedicated to the *least pairings*. As per above, just the least pairings with a frequency equal to zero eliminate millions of lotto combinations when set as a filter (or restriction in the combination-generating software). A least pairing set to minimum frequency = 2 can eliminate *all* lottery combinations sometimes. When it hits, it can generate very few lotto combinations! That's the power of positive discrimination. There are good lotto numbers and lottery pairings, as far as frequency goes. But just one bad pairing can spoil the party! By avoiding such pairing, we can eliminate a large number of lotto combinations that have a very low probability to hit the jackpot.

We generate the pairs for a *range of analysis* (named *parpaluck* for the sake of simplification) equal to 54. We expect, based on *Ion Saliu Paradox*, 13% new pairs in the next 79 – 54 = 25 lottery drawings. We can do one or two things. We enable the same file of least pairings and play the output for the next 25 draws. We can also recreate the least-pairings lottery file for each drawing separately. We play the output only for the next lottery drawing. Of course, we can combine the two and then we purge the duplicate combinations (another function in Util.EXE).

We are free to select other *parpalucks* as well. FFG offers a wide range of possibilities. Set parpaluck to N, for example (79 in the case of lotto 6/49 pairings). We know that 63% of the lotto pairs will come out in that range. We calculate now the degree of certainty for a case equal to N * 1.5 (one and a half of N); it is something like 120 in this case. Or we can just set parpaluck = 100 draws. The degrees of certainty are 78.6% and 72.3%, respectively. Seems to me, the parpaluck equal to 100 is a more efficient method of applying the least pairings as a lottery software filter. We can play the next 21 drawings and expect just under 10% of new lotto pairs to come out. We have here a 90% degree of certainty that all lotto pairs will repeat in that range of 21 drawings (from 79 to 100). It looks like a good bet to me!

Step Three: Three Numbers at a Time (Triples)

Next to the pairs are the triplets on the lottery stepladder. Every lotto number comes out mostly with two other numbers and thus forming a triple or triplet (or any similar name).

There is a caveat. The lotto triplets require a much larger lottery data file. That is, a much larger number of past drawings. Few lottery commissions have comparable lotto game histories.

How many triples or triplets are there in a lotto game, say “6 from 49”? Again, we have two easy methods to calculate. The 49 numbers can have C(49, 3) = [(49 * 48 * 47) / (1 * 2 *3)] = 18,424 triples. The lotto game draws 6 numbers; therefore total triplets in the draw is C(6, 3) = 20. In final analysis, the 6/49 lotto game yields 18,424 / 20 = 921.2 or approximately 922 integer elements.

The other method does it in one step, very much like calculating the lotto combinations: [(49/6) * (48/5) * (47/4)] = 921.2.

The formula above can be reworked to calculate triplet probabilities directly: [(6/49) * (5/48) * (4/47)] = 0.00108554 or 1 in 921.2.

We shall run the same probability and statistics software: *SuperFormula.EXE*. We need to calculate the FFG median for lottery triples. We should choose “Option 2: The program calculates p.” I suggest we use 10 for the first element of p and 9,212 for the second element. The result for FFG median (DC = 50%) is something like 638. So half of the lotto 6/49 triples will come out in 638 drawings. Wait a minute! The *WheelCheck.exe* program generated a lotto wheel that covers 3 of 6 from 49 in 514 combinations! Actually, that wheel can be reduced to 400-something lotto combinations while preserving the minimum guarantee! Well, that’s the power of reduction!

Step 3 (and further) was very poor when it came to lotto software availability. The very old 16-bit Tools.EXE (1995) lottery software calculates the frequencies of triplets for 5-, 6-, and 7-number lotto games. September of the year of grace 2008 trumpeted great news. New lotto software, of course!

I greatly upgraded the lotto utility software programs known as Util-532.exe and Util-632. I also renamed the two software packages for 5—and 6-number lotto games: *SoftwareLotto5.exe* and *SoftwareLotto6.exe*. I also made versions of the two programs *absolutely freeware*: no membership required.

The two new lotto software programs work with single and multiple number groups for 5-number lotto and lotto-6: pairs (twins), triplets (trips), quadruplets (quads), quintuplets (quints). The programs also generate lotto combinations with or without favorite numbers (from one to five favorites).

This extraordinarily powerful lotto software offers quite a bit of, well, power! Take for example a lotto 6/49 game. Eliminating the "least singles" generates 100,947 combinations (with no favorite numbers). Indeed, other least groups are even more potent (e.g., least pairs generates 13,165 combos without favorites). Enabling both least singles and least pairings generates 505 lotto combinations (down to earth from 13,983,816). Playing 2 favorite lotto numbers and eliminating the "least triples" generates only 4 combinations, sometimes only one combo!

A reasonably fast computer is needed for larger lotto groups, especially quads and quintuplets (applicable to lotto-6 only). For example, the quadruplets amount to 211,876 4-number groups or 14,125 derived elements ("4 in 6" lotto 6/49 groups). The FFG median is 9,790 drawings. I don't think there is or will be a lotto game history that long—97 years—with two lotto drawings a week!

Still, we can use those free *simulated* data files that these very applications create themselves with ease (file names in the form SIM-5 or SIM-6). You can try to generate the reports for the triplets, quadruplets, and quintuplets by using your D5 or D6 files. Again, it is not exactly like using real data files with actual lotto draws. On the other hand, the lottery commissions always run fake drawings. That is, they conduct a number of drawings before the real one (the drawing or result they publish).

Please do yourself a favor and read carefully this material presenting the powerful lotto software:
SoftwareLotto.EXE: Special upgrades to the lottery utility software for 5—and 6-number lotto games
(saliu.com/gambling-lottery-lotto/lotto-software.htm).

I know how amazed the skeptics are. Even the cynics are astonished how faithfully real-life lottery follows the laws of mathematics. When I put fundamental in the Fundamental Formula of Gambling (FFG)—I meant it. I mean it now to a degree even higher than incipiently. The least parameters created by my lottery software are very close to what FFG calculates. I generated the least files for the median (i.e., for a degree of certainty DC = 50%). Here are two cases:

The least singles file contained 26 single numbers. That is, 23 lotto 6/49 numbers repeated under the FFG median; 26 numbers did not come out. According to the fundamental formula, the ratio should be 24/25.

The percentage is even closer for higher number groups. The least pairings file contained 585 lottery pairs. It is significantly closer to the midpoint of 1,176. Total number of lotto 6-49 pairs: 1,176; half point is 588. QED.

This has been a quite-comprehensive introduction to lottery mathematics. The treatise can be expanded a lot by analyzing the multitude of filters already available in my lottery and lotto software. I am talking now about the *derived filters* or second-degree filters, as opposed to the three *primary filters* analyzed in this chapter. The derived filters aren't much different, if different at all. They also have FFG medians and degrees of certainty.

Analyzing all those filters would probably require 10 times more publishing space than this entire book! What I will try to do, however, is to present the most important lottery strategies I discovered. I will also make reference to the corresponding software I wrote. I will do my best to be as clear as possible while I must be concise.

Let's not call this chapter an introduction anymore. Let's address it properly: *Lottery mathematics in a nutshell.*

Chapter XIV

THEORY OF PROBABILITY APPLIED TO LOTTERY: STRATEGIES, SYSTEMS, SOFTWARE

1. Selecting Lotto Numbers Based on *Skips*

The lottery is so complex that it requires a big book of its own. I will try to present here the most important lottery strategies I devised based on theory of probability. I will also make an honest attempt, at least, to mention the most important lottery programs I wrote. Once again, I encourage you to still visit my Web site where you will find even more details.

I already mentioned in this book my beginnings in devising lottery strategies and systems *before* I started to write software of any kind. The manual strategies were based on *single* lotto numbers based on recent appearance. It was a painstaking process to analyze the last 40–50 lottery drawings to establish the most frequent numbers. Then, keep in the pool only numbers that had the best frequencies of *pairing* with other numbers. Only a computer program can guarantee 100% accuracy in calculations. As you read, I did have success with this type of manual strategy.

I mentioned another pioneer in this field, Gail Howard, who applied a similar pencil-and-paper method of selecting lotto numbers.

I delved more deeply into theory of probability. I discovered the very fundamental concept of *degree of certainty*. I also derived a very important parameter: *FFG median.* The lotto numbers tend to repeat more often when their running skip is less than or equal to the *FFG* or *probability median.* The probability median or FFG median can be calculated by the Fundamental Formula of Gambling (FFG) for the degree of certainty DC = 50%. This revolutionary premise constitutes the backbone of the lottery and lotto strategy that became one of the first Web pages at my site. It soon became a very popular destination.

Turning again to the 6/48 Pennsylvania lotto game, the one I analyzed in the beginning. For *DC = 50%*, the table shows the number ***6***. It means that half (50%) of the winning lotto numbers in each draw are repeats from the past ***6*** drawings! That is, on the average, *three of the six winning numbers* have also been drawn in the *last 6 lotto draws*. For *DC = 75%*, three-quarters of the winning lottery numbers in each drawing are repeats from the past ***11*** draws! That is, on the average, *four or five of the six winning lotto numbers* have also been drawn in the *last 11 drawings*.

I studied for a long time the 6/48 lotto game conducted for many years by Pennsylvania lottery. The gambling formula FFG and the numbers in the table were validated to a high degree. There were cases of lottery drawings in which *all six winning lotto numbers* had been repeats from the past 11 to 12 lotto drawings, even from the past 6 to 7 draws! In the last 500 drawings of the late *Wild Card lotto game* (6 of 48), in 123 cases (25%), all 6 winning lotto numbers were repeats from the last 12 lottery drawings. In 25 cases (5 times a year), all 6 lotto winners were also drawn in the last 7 lottery draws.

More amazingly, in 7 cases, all 6 winning lotto numbers were repeats from the last 4 to 5 drawings! Such favorable cases were separated by about 50 lottery drawings. In other words, one could select lotto numbers from the last 5 drawings and, say, win then wait 50 lottery drawings before selecting again numbers from the most recent 5 draws. The lotto wheels in my freeware program *MDIEditor and Lotto WE* offer the *4-out-of-6* minimum guarantee. They offer, however, a good shot at higher prizes. It is also true that there were situations when only 2–3 lotto numbers were repeats from the last 6 or even 11 lottery drawings.

In order to apply this knowledge to your playing strategy, you need my lottery software *MDIEditor and Lotto WE.* Run the free lottery software application. Make sure you have created a lottery drawings data file and you keep it updated. Do the statistical reporting for the lotto-6 game. Go to the skip chart and only look at the first number listed in each skip string. For example, at lotto number *13*, you will see a skip string like *4 11 9 21 . . .* The number of importance to you is 4 (the beginning of the string). It is best to work with DC = 75% since it offers a better chance to predict 4 or 5 winning lotto numbers for the next draw. It also offers a better frequency of situations when all six winning lottery numbers are repeats from the last 11 to 12 draws. Write down all the lotto numbers for which the skip string starts with a value less than or equal to the value corresponding to *DC = 75%.*

You will come up with 15–25 lotto numbers, depending on your lotto game. You will not play all the possible lotto combinations since the price of playing them consistently would be prohibitive. Instead, you will use an abbreviated lotto system or lotto wheel. *MDIEditor and Lotto WE* comes with 20-plus such lottery systems, some of the best lotto wheels anywhere. They emphasize higher prizes rather than lower costs. Say you came up with 18 lottery numbers to play for the next drawing. Click on File, Open, and select the file SYS-18.46. Follow the instructions on how to apply your lottery picks to that particular lotto wheel. My lottery software download site offers plenty of lotto wheels including for Powerball, Mega Millions, and EuroMillions lotteries.

If you came up with too many lotto numbers for your budget, you can reduce them further. Read the *Lottery Tutorial* in *MDIEditor and Lotto WE* lotto software. You may want to avoid the worst lottery number pairings in your picks. The filters in my lottery-lotto software provide countless possibilities. They are called *lottery strategies* for simplification. *A lotto or lottery strategy is a collection of filter settings applied to LotWon/SuperPower/MDIEditor and Lotto WE/lottery software*. I haven't been able to count ALL possible strategies in my lottery software!

2. SkipSystem.EXE: Powerful Software to Create Lottery Systems Based on *Skips*

I needed software to select automatically lottery numbers based on their most recent skips. The program name: *SkipSystem.EXE*. I made the software so flexible that the user had the capability to add its own parameters. For example, select the last two skips under, for example, 8 and 9, respectively. The software was debated in my forums. There were reports of extraordinary success playing online roulette. Since I consider online gambling to be fraud prone, I did not include

SkipSystem.EXE in the chapter I dedicated to roulette in this book. The system had also success in professional American football betting. You might want to read at least one topic on my message board:
http://lotterygambling.phpbbnow.com/viewtopic.php?t=75.

We'll look now, yes, at a football case:
~ the degree of certainty, DC
~ the skip corresponding to the degree of certainty
~ the cycle of fruition for the system to hit

I set the degree of certainty to be close to ***1/e*** as to be correlated to Ion Saliu's Paradox of N Trials. The degree of certainty, DC, corresponds to shorter skips. In turn, shorter skips lead to shorter cycles of fruition.

The *cycle of fruition* is a tricky parameter. I haven't found a formula to quantify it (in relation to DC or skip). But we can visualize it this way. The skip is 5, for example. If we play a system under the skip of 5, the system requires from 1 up to 5 draws (trials, spins, etc.) to complete. The ideal situation is skip = 1. The cycle of fruition is 1: the number either hits in the next drawing or it turns into a loser.

The cycle of fruition, as I see it, requires that the system numbers be played for the next several drawings. Some of the numbers will no longer meet the skip restrictions after a few draws; therefore, they must be discarded. New numbers become eligible for the system based on their new current skip. Perhaps there are better (easier) ways, but I haven't discovered them yet.

I'll present next a few recent examples or samples of my playing with SkipSystem.EXE in American football. The program generates 5 systems automatically. The reports show also the file names. For example, in the case of football, the automatic systems are named FFG-1F.SYS to FFG-5F.SYS. You can still create your own system based on the last two skips; you choose the file name.

The program treats the gambling systems based on
~ two consecutive skips and
~ three consecutive skips.

Each category, in turn, is founded on two parameters:

~ UNDER/EQUAL TO a skip corresponding to the *1/e* degree of certainty
~ ABOVE a skip corresponding to the *1/e* degree of certainty

At the end, the report shows total number of hits for each system. Actually, the number of hits counts the number of occurrences for each system in the range of past drawings (horse races, spins, football weeks) analyzed.

The program does *not* count the first skip in the string and the last skip (i.e., for each number analyzed). The method is a lot more accurate than before. The first skip is not complete yet; it is still running, and we don't know when it will end. The last skip is not certain: we don't know if the program had sufficient data to establish the last skip.

The counting of a system occurrence starts at the skip in front of the last one and goes to the beginning of the string named *Any Position.*

Football Skip Chart and System Hits
File: NFL.ODS; Date: 10-25-2006
Weeks Analyzed: 100

Team: BEARS
Any Position -> 2 1 2 1 2 1 2 1 1 2 1 1 1 4 1 5 3 1 1 5 1 4 2 1 1 2 1 1 2 7 5 1 7 2 1 1 1 3 1 2 1 1 2 2 2 2 2 1 2

* System #1 ~ File FFG-1F.SYS:
~ First 2 skips SUM-UP TO 2 -> 9 times in 49 hits (18 %)

* System #2 ~ File FFG-2F.SYS:
~ 2 skips UNDER/EQUAL TO 1 -> 9 times in 49 hits (18 %)

* System #3 ~ File FFG-3F.SYS:
~ 2 skips ABOVE 1 -> 9 times in 49 hits (18 %)

* System #4 ~ File FFG-4F.SYS:
~ 3 skips UNDER/EQUAL TO 1 -> 2 times in 49 hits (4 %)

* System #5 ~ File FFG-5F.SYS:
~ 3 skips ABOVE 1 -> 4 times in 49 hits (8 %)

* Your System Hits <= 1 1 -> 9 times in 49 hits (18 %)

****

* Total hits 2 skips SUM-UP TO 2 -> 252 hits in 100 (252 %)
* Total hits 2 skips UNDER/EQUAL 1 -> 252 hits in 100 (252 %)
* Total hits 2 skips ABOVE 1 -> 403 hits in 100 (403 %)
* Total hits 3 skips UNDER/EQUAL 1 -> 101 hits in 100 (101 %)
* Total hits 3 skips ABOVE 1 -> 211 hits in 100 (211 %)
* Total hits Your System <= 1 1 -> 252 hits in 100 (252 %)

****

The system no. 3 (“2 skips ABOVE 1”) predicted 10 teams to bet on the next week:

49ERS BILLS BRONCOS CHIEFS FALCONS JAGUARS PACKERS RAVENS TEXANS VIKINGS

The 49ERS and the RAVENS were idle, so we eliminate them from the list. JAGUARS and TEXANS played in the same game, so they canceled each other. We eliminate them from the list. The remaining qualifying teams are

BILLS BRONCOS CHIEFS FALCONS PACKERS VIKINGS = 6.

The system predicted 5 winners:
BRONCOS CHIEFS FALCONS PACKERS VIKINGS.

The system no. 5 (“3 skips ABOVE 1”) predicted 3 teams to be bet on:
BRONCOS CHIEFS PACKERS.
All three were winners!

Back to the lottery. The lottery strategy in the previous section starts with the simplest of lotto systems based on skips and pools (groups) of numbers. Only the first *skip* is considered. The player can select numbers that show the first skip being under FFG median for DC = 50% or DC = 75%.

SkipSystem.EXE not only automates the entire process. It goes further. I look at the first two or three skips (the two or three most recent skips). Instead of selecting numbers that show each of the two most recent skips less than or equal to the FFG median, I add the skips. In the previous example, lotto no. 4 started with the skips 3 and 5. Each skip is under the FFG median. One possible lottery system will select no. 4 as being playable. The lotto no. 4 is also playable because 3 + 5= 8; the result is less than the double amount of the FFG median. Double FFG median is equal to 12 for the lotto 6/49 game.

The system based on the double FFG median is clearly more frequent than a system based on each of the two skips being under the FFG median. As shown in the example above,

Number: 4
Any Position -> 3 5 17 3 7 7 10 5 3 11 1 6 6
* Automatic System <= Double FFG Median Of 12 -> 5 times in 13 hits (38%)
* Your System Hits <= 6 6 -> 3 times in 13 hits (23%)

<= 6 6 -> describes a system where the user sets the first skip to less than or equal to 6 (FFG median), and the second skip at the same level.

Number: 9
Any Position -> 4 7 3 3 4 6 3 2 21 2 2 5 7 1 11 15
* Automatic System <= Double FFG Median Of 12 -> 12 times in 16 hits (75%)
* Your System Hits <= 6 6 -> 7 times in 16 hits (44%)

The Automatic System is exactly what it says. The program selects the numbers for which the first two skips sum up to Double FFG Median (12 in this real 6/9 lotto case).

The so-called *Your System* is your creation. You create your own lottery system based on the first two skips (the two most recent).

The *Automatic System* recorded 12 hits (winning situations) or 75% of cases for lotto no. 9. *Your System* registered 7 hits (winning situations) or 44% of cases for lotto no. 9.

Hint: Numbers where the Automatic System registers 50% or better hits represent the golden eggs!

You can set *Your System* to any values you want to. The two skip values do not need to be equal. If you select values above the FFG median, your lottery system will record more hits. For example, values such as 8 and 7 instead of 6 and 6.

The software defaults to the FFG median. You don't have to fire up *SuperFormula.EXE* and calculate the FFG or the degree of certainty.

It is very important to understand that the probability is very, very slim to play a system the very next lottery drawing. That's so because the multiple skips need a *cycle of fruition*. That is, it takes several lotto drawings before we see a draw when all 6 numbers have the first 2 skips as established in a system. I checked a Pennsylvania lotto 6/49 file of 269 drawings. I saw jackpot hits, but they were separated by over 120 drawings.

I have not checked what happens from drawing to drawing. I have not written the lottery software to automate the process of checking for winners. If you are the patient type, you might want to try it manually! Go back 100 draws in your lottery data file and check the next drawings one at a time. The following is a report bloc of lottery drawings.

```
POOL-6 Winning Number Checking - With 3-hits
Group of 19 Numbers = Lotto-6 Combos: 27132
The Pool: 4 5 8 9 14 15 20 25 27 28 31 34 35 36 37 38 42 47 48

Line Drawing 6 winners 5 4 3
no. Checked Hits Hits Hits Hits

.....
5 34 35 36 37 47 48 in draw # 5 *
6 8 20 27 34 35 48 in draw # 6 *
7 20 27 28 35 38 45 in draw # 7
9 5 31 33 35 37 42 in draw # 9
10 14 15 20 24 32 38 in draw # 10
.....
27 1 9 12 31 42 47 in draw # 27
30 9 16 34 36 38 46 in draw # 30
.....
```

```
134 8 9 28 34 35 36 in draw # 134 *
139 9 13 22 27 38 44 in draw # 139
142 12 24 28 34 47 49 in draw # 142
145 5 6 26 32 35 48 in draw # 145
146 5 20 28 31 38 48 in draw # 146 *
148 20 21 22 36 37 42 in draw # 148
180 9 15 18 20 25 48 in draw # 180
215 6 11 27 28 30 35 in draw #
...
257 9 16 25 43 48 49 in draw # 257
260 3 5 11 25 33 42 in draw # 260
266 4 18 35 36 39 41 in draw # 266
269 2 15 36 37 39 48 in draw # 269

Total Hits: 4 8 37 68
```

The jackpot hits were recorded in drawing nos. 5, 6, 134, 146.

The results were poor at the beginning of the game. It takes a while for a group of lotto numbers to take over for a range of drawings. At a different point in the drawings file, other numbers will take over. The jackpot situations reflect lotto numbers with good recent frequency and higher percentages of the hit situations for the lottery system.

SkipSystem.EXE works in conjunction with the *Util*.EXE* software. The lotto/lottery utility software provides for winner checking in pools (groups) of numbers directly ("Check for Winners W," then option 2). Also importantly, the lotto/lottery utility software generate combinations from pools of numbers ("Make/Break M," then "Break strings into lotto numbers"). The software has been greatly upgraded in programs that read *SoftwareLotto.exe* (e.g., *SoftwareLotto6.exe* in the integrated package *Bright6.exe*; option "S = Super Utilities" in the main menu).

There are many lottery combinations to play by using the systems created here. That's mathematics. Some lottery players might get disappointed with the truth. Be aware that the lottery is based on gigantic odds. Many more filters are necessary to reduce the amount of lotto combinations to a playable size. I created my lotto/lottery software with that goal in mind. I came up with powerful reduction tools such as lottery filters.

The main lottery/lotto strategy and software page makes a truthful assessment of my lotto/lottery theories and software and all the other theories and programs out there. It is as honest as it can be: it's mathematical!

Now, the very important aspect of the lottery software is to generate combinations from pools of numbers. There is a group of lottery software applications I created and have offered for over a decade now. There are some relatively recent titles with the *Util* radical in the file names. The radical indicates *utilities* for lottery. There are even more recent software applications that are a lot more capable. They are named *SoftwareLotto.Exe* (for pick 3 and 4 or 5-, 6-number lotto games, plus Powerball/Mega Millions).

All the above software programs have the capability to generate lottery combinations from *pools* or *groups of numbers*. The groups of lottery numbers can be listed in files, in one line or multiple lines. For example, *SkipSystem.exe* created for you a pool of lotto numbers for the 6/49 game. The file in text format consists of one line of 12 numbers. You want to generate lotto-6 combinations from those 12 numbers. Total of lotto combinations of 12 numbers taken 6 at a time is 924. In *Util-632.exe*, you select option *M: Make/Break/Position.* Then, option *Break*, then option *2 = All 6 Numbers Equally.* The function will generate your 924 lotto combinations in a matter of seconds.

The same function can also generate lotto combinations from multiple lines of 6+ numbers each. For example, you had 49 lines, for each of your lotto 6/49 game; each line has 10 other lotto numbers as the best pairs of each of the 49 numbers. Since the lines will have common numbers, your lotto combinations will still be unique. My lottery software takes care also of eliminating duplicate combinations.

The *Position* feature is even more potent. You can generate lotto combinations based on positions or *positional ranges*. If you run the statistical functions of my lottery software (plenty of them), you will see that the lotto numbers are strongly biased regarding the position. You can read a lot about ranges or positional ranges in lotto. This is the main Web page: *Saliu.com/Newsgroups.htm.*

You will see living proof that the lotto numbers follow the Fundamental Formula of Gambling (FFG). Each position has lotto numbers based on the FFG median. Just look at the new *Powerball* game format (started January 2009). In 36 drawings, only 19 out of the 59 Powerball regular numbers have come out in the first position. The numbers 2 and 7 came out 5 times apiece; the Powerball regular numbers 1, 3, 5, 6, 8 came out 3 times each. Meanwhile, numbers 24 to 59 have not hit yet. For the fifth position, only 18 numbers have hit, especially in the range 50 to 59. You can use the same option, *M: Make/Break/Position*, in *Util* software but select *Position/Ranges*.

The previous format of the Powerball game drew 5 regular numbers from a field of 55 and one Powerball from a field of 42. *SkipSystem.exe* registered some spectacular results! The system hit at least 4 (four) Powerball jackpots as of August 18, 2007 (in a 20-drawing span: draw nos. 3, 8, 9, 20).

3. The *Wonder Grid:* Lottery Systems Based on *Pairings*

The following lottery strategy was first published in 2001. It was derived from my very first lotto strategy when I selected frequent numbers and their best pairings. The strategy is based on two elements:

1. Playing a *favorite* (*key*) number based on the Fundamental Formula of Gambling (FFG)
2. Playing only the *most frequent pairings* of the favorite number

Selecting a *key* (favorite) number follows the procedures presented on the Winning Lottery Strategy page (Saliu.com/LottoWin.htm). There is no denial at this step. Each and every lotto number repeats after a number of drawings less than its median in at least 50% of the cases. FFG explains in mathematical terms such behavior of lottery numbers. The formula is also validated by real lottery drawings. I haven't heard of any report to the contrary in any lottery. I am convinced now FFG will not be contradicted in this regard, ever.

Let's take the lotto 6/49 game, the most common worldwide. The median is 6 for the 50% degree of certainty. We need software to plot the skips for each lotto number. A skip shows how many drawings a number waits between hits. My freeware *MDIEditor and Lotto WE, Util-6.EXE, SoftwareLotto6.exe* (possibly more titles!) do the best job at charting the lotto skips. The skips will look like this:

Number: 8
* Skips: 4 11 20 1 22 6 4 7 3 8 4 3 2 6 0 12 13 0 6 1 3 1 3 25 6
* Median Skip: 4

Number: 9
* Skips: 5 5 2 0 35 1 1 4 0 23 1 11 16 5 0 17 21 11 4 3 8
* Median Skip: 5

If we add up the skips 0, 1, 2, 3, 4, 5—for any number—the sum will represent at least 50% of all cases. (The case above shows actual data in a 6/69 lotto case, where the median is 8.) Calculating the probability to selecting a lotto number with a skip below the median involves two steps. There is a 50% chance (*1/2*) that a lotto number will repeat after an amount of drawings lower than the median. But the median takes *6* values from 0 to 5. The simultaneous probability of the two events is *1/2 x 1/6 = 1/12 (8.3%)*. We should be successful every 12 tries (drawings) in picking a favorite lotto 6/49 number.

The second part of the strategy is based on the *pairings* of the lotto numbers. Given a range of drawings, each lotto number shows a clear bias toward being drawn with the rest of the lotto numbers. UTIL-6.EXE calculates the frequencies of all lotto pairings for every lotto number. The report looks like this:

Number: 1 Hits: 19 (9.5 %)
With #: 8 16 30 22 25 14 31 32 42 27 29 17 21 15 33 35 36 39 5 44 51 52 . . . 50 26 12 28 13 56 57 4 61 62 2 64 65 6 69 24 59 60 38 11 46 34 55 66 67 19 20
Hits: 4 4 4 3 3 3 3 3 3 2 2 2 2 2 2 2 2 2 2 2 2 . . . 1 1 1 1 1 1 1 1 1 1 1 1 1 1 1 1 0 0 0 0 0 0 0 0 0 0 0 0 0
Pairs total: 95

Number: 2 Hits: 23 (11.5 %)
With #: 36 40 6 48 68 24 25 7 9 42 11 50 61 6 . . . 69 41 10 43 5 45 22 23 12 8 26 52 15 56 28 30 31 32 63 34 16 17 18 39 1 4 37 3 62 33 53 46 55 29 57 14
Hits: 6 6 4 4 4 3 3 3 3 3 3 3 3 3 3 3 2 2 2 . . . 1 1 1 1 1 1 1 1 1 1 1 1 0 0 0 0 0 0 0 0 0 0 0 0 Pairs total: 115

Number: 69 Hits: 18 (9 %)
With #: 19 51 63 3 22 26 36 40 6 9 64 29 34 . . . 45 5 50 20 21 55 56 57 10 60 62 23 24 65 11 67 68 41 28 35 7 58 30 46 61 17 48 49 32 33 52 53

Hits: 4 4 4 3 3 3 3 3 3 3 3 2 2 2 2 . . . 1 1 1 1 1 1 1 1 1 0 0 0
0 0 0 0 0 0 0 0 0 0 0 0 0
Pairs total: 90

The *top 10%* of the pairings represents *25%* of total frequencies for each number. The top is defined as the most frequent part in a pairing string (starting from left to right). In the case above (still a lotto 6/69 game), the top 10% means the first 7 pairs; 10% of 69 is approximately 7. The first 10% pairs for lotto number 1 add up to 24 (25.3% of all 95 cases for that lotto number). It's 4+4+4+3+ 3+ 3+ 3 = 24.

Conversely, the bottom 10% pairs sum up to 0 (zero).

The top 25% of the pairings represents 50% of total frequencies for each number. That is, the 25% segment of the most frequent pairs account for 50% of the entire pairing frequency for any given lotto number.

The top 50% of the pairings represents 75% of total frequencies for each number. That is, the first half of the pairs (from left to right) accounts for 75% of the entire pairing frequency for any given lotto number.

Based on the percentages above, I decide

1. to play a favorite lotto number, AND
2. to play only its top 25% pairs.

In a hypothetical case, I will play number 1 since it shows as the last skip a value between 0 and 5. For a lotto 6/49 game, I will also play the most frequent 12 pairs (25% of the remaining 48 lotto numbers). Lotto number 1 comes out with its top 25% pairs in 50% of its appearances. The chance is 1/2 that the lotto number will come out with such pairings—*when* it hits!

Every combination I play will contain the number 1 (the favorite). The remaining 5 numbers will consist of combinations of the 12 numbers in the most frequent 25% pairs. Total combinations of 12 taken 5 at a time: C(12,5) = 792. The combined probability to hit the jackpot is 1/12 x 1/2 = 1/24.

It sounds extraordinary. The chance is 1 in 24 to win the jackpot with 792 combinations: 792 x 24 = 19,008. For a total cost of $19,008, it is possible to win the one-million-dollar jackpot! Actually, I saw such lotto combinations containing a favorite number plus 5 numbers from its top 25% pairs. Visitors to my site also noticed such favorable occurrences. Checking for winning lotto numbers this way is a little trickier. First, the software should break down the top pairs in groups of 5 and then add the favorite number to each combination. Then it should check the results against real drawings.

I named *wonder grid* or *lottery wonder grid* a set of combinations consisting of each lotto number and its top pairings (pairs). For example, each lotto 6/49 number is paired variably with the other 48 numbers. The frequency of the pairs goes from 0 (no show) to 5 or more in 100 draws. A wonder grid will be constructed for each lotto number and its 5 most frequent pairs. For example,

1 20 43 8 10 37
2 23 28 31 34 9
. . .
49 42 10 4 44 16
(49 6-number combinations to play)

I have done extensively more research in the field after 2003. In the lotto games, the top 4 (lotto-5) or top 5 (lotto-6) do not hit in the next draw in many, many thousands of draws or simulated combinations. The top 1-2-3-4-5 pairs may hit within a range of future draws, if the same wonder grid is played. My free software *GridCheck632.EXE* shows the winning strings and their skips for a tough lotto 6/69 game. The wonder grid does not hit in the next very few draws. I discovered that consecutive pairings such as 1-2-3-4 or 1-2-3-4-5 are no-shows! Also, the pairings from draw to draw show real randomness in the line of the draws themselves. Here is a sample report for Pennsylvania lotto 5/39 game. The report shows each number in the draw and its pairings. For example, the first number in the draw came out with its pairs of indexes 4-18-21-23. No 1-2-3-4!

The analysis shows, however, that top pairings come out reasonably frequently. Instead of pairs 1-2-3-4, the range of pairs from 2 to 9 should be selected. They had jackpot hits in draws no. 6 and no. 63 (the pairings marked with an *). Three of the pair indexes are consecutive, but not all four. Also, the second report shows that none of the pairings was a repeat. The top 8 pairings had also several 4 winning numbers, when only three of the pair indexes were correct.

```
LOTTO-5 Pairing Report by Draw

File: LOTTERY\LOTTO-5\PA-5

Draws Analyzed: 1000

Pairings Draw Range: 78


Line Num Num Num Num Num

no. 1 2 3 4 5


1 4 18 21 33 7 13 28 36 14 18 21 22 2 18 24 26 15 19 20 36

2 12 17 26 35 9 14 23 36 23 25 35 36 3 10 11 29 17 19 25 27

3 14 15 21 37 14 21 24 25 20 25 26 37 19 20 36 37 17 18 19 28

4 8 16 17 30 13 22 30 33 18 19 25 27 15 22 23 33 21 25 32 37

5 15 16 28 32 3 16 28 38 1 8 22 32 4 8 27 32 1 5 6 36

6 * 2 6 7 8 7 20 22 32 12 22 30 36 20 29 30 34 5 12 26 27

. . .
```

```
53 5 7 8 19 2 10 21 23 1 6 28 30 11 12 16 21 18 23 24 28

61 4 9 23 33 6 7 8 10 3 6 9 22 5 20 23 37 5 12 21 25

62 1 3 7 19 5 7 17 34 4 11 15 29 4 7 16 17 3 15 27 32

63 5 8 22 26 4 10 35 36 14 27 33 37 * 2 7 8 9 9 11 25 38

64 23 26 33 37 2 17 24 26 15 29 32 34 11 17 22 25 1 19 26 34
```

How to apply the wonder grid theory:

1. Follow the tactic of *GridCheck632.EXE* (explained later).
2. Create a wonder grid for pairs 2-9 (lotto-5) or 3-12 (lotto-6). Do not select all consecutive indexes for a wonder grid. Do not select a repeat pairing string. Eliminate some combinations using the skips as filters. Allow a wait period between hits.
    2.a. If you select pairings 3-6-7-9-10, create the grid every time you play; the combinations may change from draw to draw.
    2.b. You can play the same grid for the next draws without changes.
3. You can select more than one pairing strings (for example, 3-6-7-9-10 and 3-5-7-8-11 or more) and eliminate combinations using the skips or standard LotWon software filters.

Applying the concept to the pick-3 game, the behavior of the wonder grid is different. Playing the best 3 pairs while eliminating the worst 7 pairs is one proven strategy at beating not only random play handily, but also beating the monstrous house edge of 50%! I create the final grid and play it for the next 100 draws. This appears to be the optimal range for the fruition of a pick-3 wonder grid. There are two interesting triggers:

1. If the final grid consists of 18 or 19 combinations, it is best to play the same grid for the next 100 draws.

2. If the grid consists of 10 or 11 combinations, the probability is higher, and the grid will hit the very next draw. If the grid does not hit, discard of it.

Here is an exemplification of playing case scenario 1.

```
PICK 3 Winning Number Checking
Files: OUT3 (19) against > PA-3 (100)

Line Combination Straight Boxed
no. Checked Winner Winner

1 0 4 9 in draw # 22
1 0 4 9 in draw # 77
2 0 9 4 in draw # 22
2 0 9 4 in draw # 77
4 1 1 6 in draw # 2
4 1 1 6 in draw # 67 *
6 1 6 1 in draw # 2 *
. . .
16 8 1 6 in draw # 93
17 8 6 1 in draw # 93
18 9 0 4 in draw # 22
18 9 0 4 in draw # 77
19 9 4 0 in draw # 22
19 9 4 0 in draw # 77

Total Hits: 5 19
```

I get the same number of straight hits consistently, for either 18 or 19 combinations. No other strategy is involved. Even if playing the same grid every draw for the next 100 draws, COW or the cost of winning is 1,800 (1,900). The five wins amount to 2,500. The result is profit. Not only does this strategy beat random play, it also beats the monstrous house edge or house advantage imposed by the lottery (50%)! Random play will yield 1.8 straight wins (1,800/1,000).

Of course, the skips are omnipresent. Nothing can avoid skipping—it's mathematical! The strategy above does not hit immediately in most situations. The player can safely sit out 5 drawings between hits. But this pick-3 pair strategy should hit within the next 12–13 draws, based on my FFG calculations. The hit in draw no. 93 in the report above indicates a hit after 7 drawings (100 – 93).

I prefer changing the pairing sets after the first hit. I redo the wonder grid at the winning drawing. This is a very favorable case. I skipped 5 draws; I paid for 2 draws: $38. I won straight. I discard of the strategy. If I continued to play, I would skip 5 drawings again; 93 – 77 – 5 = 11. 18 * 11 = $198. Still a good outcome.

If reversing and *playing the worst three pairs while eliminating the top three pairs* does not yield any straight wins for most cases. Also, there are significantly fewer boxed hits. Unquestionably, the top pairings show a significantly higher frequency in regard to future draws.

These discoveries are consistent with the Fundamental Formula of Gambling (FFG). Each number has a strong tendency to repeat after a number of trials less than or equal to the FFG median. Pairs of numbers have necessarily the same tendency. Pairs of numbers that came out together have then the tendency to repeat more often for ranges shorter than the FFG median. The clear trend continues for a number of future trials. I noticed, in fact, that the trend goes both directions: future and past. I am not talking about the past as in curve-fitting cases. I usually do a pairing for 20 draws (pick 3). I check the past *beyond* the 20 draws analyzed. The past 100 draws show a trend similar to the one for the future 100 draws.

Perchance that *GridCheck632.EXE* sounded mysterious! The program checks the past performance of the *lotto wonder grid* for the 6-number lotto games. The software automatically creates three files (named GRID6) for various draw ranges (*parpalucks)*. The program starts a number of drawings back in the DATA-6 file and creates 3 GRID6 files: for N*1, N*2, N*3 draws. *N* represents the biggest number in the game. By default, the program checks how the GRID6 files performed in the previous 100 draws (the span). Your data file must have at least {SPAN + (N * 3)} draws. If your lotto-6 game has 49 numbers and SPAN=100 (check the previous 100 draws), then your files must have at least 100 + 49 * 3 = 247 drawings. The program will let you know what to do to overcome the error.

The software creates the three GRID6 files in the background. The first one is a range equal to the biggest number in your lotto game (e.g., 49). The second file is created for a range equal to N*2 (the biggest lotto number times 2). The third file is created for a range equal to N*3 (the biggest lotto number times 3). The GRID6 checking reports will be saved to three disk files. Defaults are ChkGrid6.N1, ChkGrid6. N2, ChkGrid6.N3. They show how many hits (if any) the corresponding wonder grid had for a particular drawing. Here is a more clarifying example, with data for Pennsylvania 6/69 lotto game (now discontinued).

```
GRID6.N2 Winning Lotto Number Checking

~ Lotto Draw # 37 > 2 4 8 11 16 53
~ Files: GRID6.N2 ( 69 ) against > PA-6 ( 37 )

Line Combinations 6 5 4
no. Checked Hits Hits Hits

16 16 2 3 40 49 32 in draw # 11

Total Hits: 0 0 1

~ Draw # 38 > 1 9 31 45 55 59
~ Files: GRID6.N2 ( 69 ) against > PA-6 ( 38 )

Line Combinations 6 5 4
no. Checked Hits Hits Hits
```

```
1 1 10 31 26 11 22 in draw # 32
16 16 2 3 40 49 32 in draw # 11
22 22 1 31 32 41 59 in draw # 32

Total Hits: 0 1 2

~ Draw # 39 > 18 21 23 25 43 53
~ Files: GRID6.N2 ( 69 ) against > PA-6 ( 39 )

Line Combinations 6 5 4
no. Checked Hits Hits Hits

1 1 10 31 26 11 22 in draw # 32
16 16 2 3 40 49 32 in draw # 11
22 22 1 31 32 41 59 in draw # 32

Total Hits: 0 1 2

~ Draw # 40 > 7 8 19 36 40 45
~ Files: GRID6.N2 ( 69 ) against > PA-6 ( 40 )

Line Combinations 6 5 4
no. Checked Hits Hits Hits

1 1 10 31 26 11 22 in draw # 32
16 16 2 3 40 49 32 in draw # 11
22 22 1 31 32 41 59 in draw # 32

Total Hits: 0 1 2
```

The lotto draw range or the span of analysis is the most important element. I'm not sure about the optimal value of the range. First, I said it was N*3; then I said it must be N. It appears that the best range is N*2. The range N*1 is also a good value. As of *three-times-the-biggest-number (N*3)*, it probably creates an outdated grid file. A range beyond *N*3* is useless. You can look at the three reports and see which one(s) fared the best in your lotto-6 game. Again, I'm (almost) convinced that the best ranges are N*2 followed by N*1. The case above is the report for an N*2 grid file.

1. The program started at draw no. 2 and created a grid file for a range of 69 * 2 = 118 draws. Therefore, the lotto software created the first *wonder grid*: the best *69 pairings* in the range 2–120 past draws. The program stopped at draw no. 120 in the Pennsylvania data file. Next, the lotto program created a winning report for draw no. 1. Important fact: the program does not create a *lotto wonder grid* for the current draw. It would be a case of *curve fitting*: the wonder

grid would hit the jackpot every ten draws or so! But such case would not take place in reality! We must exclude the current draw from the range that creates the wonder grid. The winning report (almost the same as in *Util632.EXE* or *Winners.EXE*) checks for future drawings. In this first step, the future consists of one draw only: no. 1. The winning report checks for "6 of 6" winners, "5 of 6" winners, and "4 of 6" winners. I excluded the "3 of 6" prize from the reports. There are way too many lines! The reports would be too large and slower! Right now, it takes around one minute to generate the lotto *wonder grid checking software*. Usually, there are no hits at position no. 1.

2. The second step. The program continues at draw no. 3 and creates a grid file for a range of 69 * 2 = 118 draws. Therefore, the program created the second wonder grid: the best 69 pairings in the range 3–121 past draws. The program stopped at draw no. 121 in the PA-6 data file. Next, the program created a winning report for draws no. 1 to no. 2. In this second step, the future consists of two draws: no. 1 and no. 2.

Usually, the wonder grid starts hitting after 5–6 draws ("4 of 6"). Higher prizes require a wider gap between hits. In the case above, draw no. 38 registered one "5 of 6" hit. The streak continued uninterrupted to draw no. 53. Had I decided to play just before draw no. 38, I would have hit 5 winners after 6 draws. That is, draw no. 32 would have given me one ticket with 5 winners (also one ticket with 4 winners). Had I decided to play just before draw no. 53, I would have hit 5 winners after 53 – 32 = 21 draws. That is, draw no. 32 would have given me one ticket with 5 winners (also one ticket with 4 winners). A good strategy would be to wait for 5 draws, then start playing for the next up to 35 draws (N/2). The skips are different from game to game. That's why the three grid reports are so useful.

How to figure out the validity and strength of various lottery strategies? The most common rule is to compare a strategy to random play. *The benchmark is the normal probability rule: 99.7% of the successes will fall within 3 standard deviations from the expected (theoretical) number of successes.* If a strategy beats the random expectation by more than 3 standard deviations, it surely has a solid foundation. The margin of error is less than 0.3%.

The game analyzed was Pennsylvania 6/69 lotto. The probability of hitting 6 of 6 is 1 in 119,877,472. The probability of winning 5 of 6 is 1 in 317,136. The lottery software generated 10,000 combinations around the FFG median. I checked 350 real draws in the 6/69 lotto game. Therefore, the number of trials was 350 * 10,000 = 3,500,000.

The normal probability rule applied to the "6 of 6" case is summarized by *SuperFormula.EXE* as follows:

The standard deviation for a lotto event of probability p = .00000001 (1 in 119,877,472) in 3,500,000 binomial experiments is BSD = .17.
The expected (theoretical) number of successes is 0.29.

Based on the Normal Probability Rule:

- 68.2% of the successes will fall within 1 Standard Deviation from 0.29 (i.e., between 0.12–0.46)
- 95.4% of the successes will fall within 2 Standard Deviations from 0.29 (i.e., between 0–0.63)
- 99.7% of the successes will fall within 3 Standard Deviations from 0.29 (i.e., between 0–0.80)

The lottery strategy I presented had one "6 of 6" winner. It clearly beat the random expectation by one extra standard deviation!

The "5 of 6" case:

The standard deviation for a lotto event of probability p = .00000315 (1 in 317,136) in 3,500,000 binomial experiments is BSD = 3.32.
The expected (theoretical) number of successes is 11.

Based on the Normal Probability Rule:

- 68.2% of the successes will fall within 1 Standard Deviation from 11 (i.e., between 8–14)
- 95.4% of the successes will fall within 2 Standard Deviations from 11 (i.e., between 5–17)
- 99.7% of the successes will fall within 3 Standard Deviations from 11 (i.e., between 2–20)

The lottery strategy I presented had 75 "5 of 6" winners. It beat the random expectation handily by 16 extra standard deviations (75 – 11 = 64; 64 / 3.32 = 19.27 standard deviations; 19.27 – 3 = 16.27 standard deviations beyond the level 3 required by the normal probability rule)!

The wonder grid consists of 69 combinations for a lotto 6/69 game. The wonder grid is expected to hit "5 of 6" within 69 / 2 = 35 draws. In this case, number of trials is 69 x 35 = 2,415.

The standard deviation for a lotto event of probability p = .00000315 (1 in 317,136) in 2,415 binomial experiments is BSD = 0.09.
The expected (theoretical) number of successes is 0.008.

Based on the Normal Probability Rule:

- 68.2% of the successes will fall within 1 Standard Deviation from 0.008 (i.e., between 0–0.098)

- 95.4% of the successes will fall within 2 Standard Deviations from 0.008 (i.e., between 0–0.188)
- 99.7% of the successes will fall within 3 Standard Deviations from 0.008 (i.e., between 0–0.278)

The lotto wonder grid also beat random expectation by 8 extra standard deviations (1 success – 0.008 = 0.992; 0.992 / 0.09 = 11 standard deviations; 11 – 3 = 8 standard deviations beyond the level 3 required by the normal probability rule)!

4. Lottery Strategies Based on Eliminating the *Least*: Number Groups with the Worst Frequency

This is my latest in lottery research and software writing. The main points were presented in the section dedicated to *lottery mathematics*.

I created very potent software that generates frequency reports for every group of numbers, single—and multiple-number groups. For example, *SoftwareLotto6.exe* performs many, many tasks for the 6-number lotto games. The program does also this: it generates frequency reports for single numbers (number by number), two-number groups (pairs), three-number groups (triples), four-number groups (quadruples), and five-number groups (quintets).

There are three reports for each group:

1. All possible groups in the game, from the first one to the last one, in lexicographical order (e.g., all 1,176 pairs [in 6/49 lotto] from 1-2 to 48-49; the frequency of each pairing is shown).
2. The groups are sorted in descending order, from the most frequent to the least frequent.

3. One report is named Least (e.g., Least62 showing the 6-49 pairs with the worst frequency, usually 0 appearances).

The *Least* groups do not come out too often to be sure. That's why their frequency is so low. No doubt, the least groups will come out from time to time. Those will be losing moments for us. For we base our strategy on ***eliminating the Least groups of numbers***. That's the concept I used, *Least file*, when I started my lotto software back in 1988. I stick with old friends!

The new *SoftwareLotto.exe* offers quite a bit of power to the user! Take for example a lotto 6/49 game. Eliminating the least singles generates 100,947 combinations (with no favorite numbers). Indeed, other least groups are even more potent (e.g., least pairs generates 13,165 combos, without favorites). Enabling both least singles and least pairings generates 505 lotto combinations (down to earth from 13,983,816). Playing 2 favorite lotto numbers and eliminating the least triples generates only 4 combinations, sometimes only one combination!

This type of software handles a multitude of lotto and lottery games: 5-, 6-, 7-number lotto; pick 3, pick 4; Powerball/Mega Millions; plus horse racing trifectas.

I will analyze here only the new functions: the *Rundown* functions and the *Combination-generating* modules. They were not presented in the superseded packages named *Util*.exe*.

1. The *Rundown* functions perform statistical analyses of single and multiple number groups in every game. The groups are

~ Singles, 1 lotto number at a time;
~ Pairings, 2-number lotto groups;
~ Triples, 3-number lotto groups;

~ Quadruples, 4-number lotto groups;
~ Quintuples, 5-number lotto groups (applicable to the lotto 6/7 games only); and
~ Sextuples: 6-number lotto groups (applicable to the lotto-7 games only).

You already saw the statistical reports for the triplets and the quadruplets in the *total freeware* (the *ToolsLotto.exe* programs). The programs determine the span of analysis or parpaluck for each number group. The span of analysis is calculated by the Fundamental Formula of Gambling (FFG). It represents N (number of past lottery drawings) for a degree of certainty, DC = 50%.

The triplets, quadruplets, quintuplets, and sextuplets require very, very large data files, especially for the lotto-6/7 games. Virtually no lottery has ever conducted that many drawings. It would be great to have real draws. Still, we can use those free *simulated data files* (named SIM*) that these very applications create themselves with ease. You can try to generate the reports for the triplets, quadruplets, and quintuplets by using your D5, D6, or D7 files. Again, it is not exactly like using real data files with actual lotto draws. On the other hand, the lottery commissions always run *fake* drawings. That is, they conduct a number of drawings before the real one (the drawing or result they publish).

Virtually, the source code in my software is optimized to the maximum. Still, the specific rundown reports in these two programs are very demanding. The quadruples, quintuples, and sextuples might take quite a long time on a slow PC. There is no way around patience in this matter!

2. Let's look now at the *modules that generate lotto combinations*, with or without favorite numbers, eliminating the least groups or not.

This new combination generating function has 5/6/7 subroutines:

2.1 Generate lotto combinations with NO favorite numbers
2.2 Generate combinations with ONE favorite lotto number
2.3 Generate lotto combinations with TWO favorite numbers (a favorite lotto pair)
2.4 Generate combinations with THREE favorite lotto numbers (a favorite triple)
2.5 Generate lottery combinations with FOUR favorite lotto numbers (a favorite quadruple)
2.6 Generate combinations with FIVE favorite lotto numbers (a favorite quintuple—for lotto 6, 7 only)
2.7 Generate combinations with SIX favorite lotto numbers (a favorite sextuple—for lotto-7 only)

Each subroutine has its own multiple choices:

2.21 Do not eliminate any least frequent groups
2.22 Eliminate the least singles
2.23 Eliminate the least (worst frequent) pairings
2.24 Eliminate the least triplets
2.25 Eliminate the worst quadruplets
2.26 Eliminate the worst quintuplets
2.27 Eliminate the worst sextuplets

Be mindful of synchronicity too. You may choose a favorite number and try to generate combinations by eliminating the least singles. If your favorite is among the entries in *Least61*, the software will not generate one single-lotto combination, of course! Also, the skips and streaks have an important role to play. The software to generate the winning reports is still on the drawing board! Perhaps, on a future edition of this book.

Yes, it can be done manually. Here is what I do. Start somewhere 100 draws back in my lotto data files. If there's not enough real lottery drawings, I use the D(5, 6, 7) data files, which include *simulated* lotto combinations. I generate all the *Least* files with *0* as the upper limit. I generate combinations by enabling the respective *Least* feature (e.g., Least singles). The amount of lotto combos generated is quite impressive. I use the *Check for winners function* (in the *SoftwareLotto.exe* programs). I apply the winners function for the next 100 draws (that behave here as future lottery drawings). I see how that particular *Least elimination* feature fared. I go back and set the upper limit to 1 and follow the procedure of generating lotto combos and checking them against the next 100 lottery drawings. I go back again and set the upper limit to 2 and follow the procedure above. Set the upper limit to 3 and so on.

Checking for lotto winners allows me to see a *cycle of fruition* or *potrocel* (name chosen for the sake of simplification). I explained in the lottery mathematics section that the degree of certainty comes to fruition within a range of future trials (lotto drawings, in this case). Furthermore, the *potrocel* helps with figuring out the *median skip* of the respective strategy.

It is really tedious! I want the software to do that automatically for me. It should work like the winning report generators: the lottery software that creates the WS files (e.g., W6.1, MD5.4, etc.). I want the software to tell me, "The upper limit = 3 for lotto 6 is about to hit in the next few lottery drawings."

I do know that high upper limits (especially for singles and pairs) are quite frequent—even 4 or more. Such *Least* files absolutely obliterate the lotto odds! Since they don't hit frequently, they don't generate lotto combinations most of the time. But they do hit—with just one lotto combination sometimes.

I can increase the winning probability even further. I generate the *Least* files for several past drawings, usually adjacent ones. For example, I go back 6 drawings, then 7, then 8. I combine all output files and play them for the *potrocel* (as calculated by FFG). Right now, I can only play blindly, but having a high degree of certainty that I can win big, really big, in a reasonable time frame. On the other hand, I do not do things before I know as precisely as possible the underpinning mathematics of things. And there you have the reason . . . I am referring to those who can't comprehend *why I do not posses* all those millions and millions of dollars in lottery winnings!

5. Lottery Strategies Based on the *Birthday Paradox*

There is one more type of strategy I want to present to you *before* we reach the *filtering* concept. We talked a lot here about the *Birthday Paradox* or *probability of repetition* or *duplication.* The principle *can* be applied to gambling and lottery. The *Birthday Paradox* or *probability of repetition* stresses the role played by the *number of trials*. In other words, the *past counts*.

For example, I look at a sequence of 8 roulette numbers as an eight-element string. The degree of certainty is better than 50% that such string should contain one repetition (duplication). One of the eight numbers should be a repeat with a fifty-fifty chance. The same is true about lottery drawings. In this case, the element is the index of the combination drawn. Every lotto combination, for example, is defined by an index or lexicographical order or lexicographic rank.

With this new knowledge in mind, I studied some real data: lottery drawings and roulette spins. I was somehow surprised to discover that repetition occurs close to that cutoff point of the fifty-fifty chance (DC = 50%)! I should also point out that the strength of the analysis and system creation is stronger at the beginning of the game. For lottery and lotto, the beginning is clear: a game starts with the first draw of a game format. For roulette, a beginning is the very first spin of the day.

My mathematical analysis has been helped a great deal by new games in Pennsylvania State Lottery. The state lottery commission started a new pick 5 digit lottery game named Quinto. It draws 5 digits from 0 to 9: 00000 to 99999 (a total of 10,000 possibilities). At the time of the analysis, the Quinto game consisted of 397 drawings. I used my own software for analysis. I included that type of software only in the horse racing bundle: *BrightH3.exe*. The name of the program goes as *UnderOverH3.exe*. I got similar software for most lottery and gambling forms. I didn't include the programs in my software bundles because I wasn't a strong believer in systems derived from the probability of repetition (*birthday paradox*). I discovered on the month of September, of the year of grace 2009, that repetition plays a very important role in gambling systems.

I looked at the report for the Quinto game. I ran *Collisions.EXE*, option *R = Reversed Duplication Problem*. The parameters are 0 and 99999 and a degree of certainty, DC, = 50% (do not type the percentage sign). The program calculates that some *373* Quinto drawings are necessary to reach a *fifty-fifty* chance of repetition. Question is should I wait 373 draws before I play the previous draws? I think of that margin of error the pollsters apply in political polling (especially during electoral seasons). It is *3%*, plus or minus. In this case, 3% of 373 is something like 11 drawings. Thus, I start playing the system 11 Quinto draws before 373. In the case of Pennsylvania lottery, Quinto, the first repeat occurred right on the nose: draw no. 373! Play *373* tickets for *11* drawings for a cost of *$4,103* and win *$50,000*!

Pennsylvania lottery has also a new 5-number lotto game (5 from 43). The game had 604 draws at the time of analysis. The *fifty-fifty* chance of a repeat combination leads to a result of *1,156* draws, with a 3% margin of error equal to *35*. Thus, the system should be played starting at draw no. 1,156 – 35 = *1,121*. It's less than three more years' worth of drawings.

How about Powerball (current format: 5 from 59 *and* 1 from 39)? That game requires 16,453 drawings as the waiting period (minus 494 draws as the margin of error). A long, long way to go (over a hundred years from now).

I looked also at roulette. I believe I got an accurate data file from Hamburg Spielbank (Casino), Germany. The file contains all the roulette spins for the month of January 2006. I will publish here a fragment of the *Under & Over* report for this roulette case. The *fifty-fifty* repeat chance for a single-zero roulette game is 8 spins (numbers). The margin of error is 1. To that effect, I wait 7 spins after the roulette table is open and becomes active.

```
* Roulette Winning Pattern Under/Over *

* File: HAMB0106.WH1


Line Spin Skip_1 Skip_2 << <> >< * >>

. . .

7934 16 7 10---+

7935 1 5 18---+

7936 27 27 2-+--

7937 2 12 19---+

7938 4 27 26 +---

7939 22 12 25---+

7940 31 27 24-+--

7941 1 27 12-+--

7942 16 3 2---+

7943 29 0 15---+

. . .
```

The spin at the bottom (7,990) represents the first spin on the month of January 2006. It is also the first recorded day of the month. Just the beginning of a day is sufficient for roulette systems. It is a fresh start—I consider it a new game. I wait 7 spins and then start playing the roulette numbers of the previous 8 spins. The next two spins are not winners, but the third one is. I paid for 8 + 8 = 16 numbers, and I won 36. I made a profit, and I might quit as well. I don't quit because I know so well the Fundamental Formula of Gambling (FFG). The numbers tend to repeat after a number of trials (spins) equal to or less than the FFG median. *SuperFormula.exe* calculates the FFG median for 8 spins to be equal to 5. It means that I can decide to play this birthday paradox system only for 5 spins following a hit. I lost the first situation (the next hit occurred after 13 spins). Then, I paid for 1 + 2 + 1 + 1 + 4 + 5 + 2 + 5 = 17 spins. I add also the first 7 spins of initial waiting for a total of 24 betting units. Total wins: 11 * 36 = 396 betting units or a profit of 372 betting units. Just a few hours in the day!

It is very important to know the start of the game. Then only play for a number of trials under the FFG median. Yes, it is likely you will not cash in all winning situations, but you keep the cost down. *Collisions.EXE* makes it very easy to do the calculations. You only need to know the total elements (sets, combinations, numbers, etc.) in the game.

For example, a lotto 6/49 game has a total of 13,983,816 combinations. In *Collisions.EXE*, select *R = Reversed Duplication Problem.* Then type

1
13983816
50

This section answers also one of those heated questions and debates. Will a lotto 6/49 combination repeat? Some say

Never! Others will say, *It will repeat forever!* You and I know it better: the degree of certainty of any repetition depends on the number of trials.

6. Lottery Strategies Based On Frequency Groups, Skips as Filters, Decades, and More

We are clear now that the lottery numbers have the same probability of being drawn. Yet we always see different lottery numbers with different frequencies. We also discovered that the numbers tend to repeat more often under certain circumstances, within certain ranges of drawings. Furthermore, the numbers tend to appear more often grouped with certain numbers. And all these tendencies are quite different from lottery game to lottery commission!

I had noticed that many lottery players and system developers believe that most numbers always come from a group of so-called *hot* numbers! No matter what, they strongly advise to play only lottery numbers from the hot group. By *hot* numbers, they understand the numbers with the highest frequency. By and large, the authors do not relate the frequency to the range of drawings analyzed.

In my book, the range of drawings always matters. It represents the *number of trials*, *N*, one of the three essential elements of the Fundamental Formula of Gambling (FFG). I went further and discovered that the lottery numbers tend to not only come out preponderantly from a group of *hot* numbers, but also from two other groups of numbers. There is a group of numbers that show a midlevel frequency; they call such group *mild*. There is also a group of numbers that show up poorly; they have low frequencies. Players and authors call this group *cold*.

I take as an example a lotto 6/49 game. We divide the lotto numbers in three frequency categories as follows:

1. Group 1, the most frequent: *1/8* (12.5%) of all numbers = 6 lotto numbers (*hot*)
2. Group 2: *37.5%* of all numbers = 18 lotto numbers (*mild*)
3. Group 3, the least frequent: *50%* of all numbers = 25 lotto numbers (*cold*)

The division has a good foundation: fewer numbers with higher frequency lead to better playing efficiency.

The 6 winning numbers in a particular lottery drawing will have a distribution such as 2-2-2: *2* numbers from the top 6, 2 numbers from the middle-frequency zone, 2 numbers from the least-frequent group.

I created a table for a few examples of distribution. The table shows how many numbers in each of the three frequency groups; number of winners in each group; then, in the brackets, how many combinations that group generates; last, how many total lotto combinations by the distribution (e.g., 2-2-2 generates 688,500 combinations for 6/49 lotto).

We can use these formulas to calculate the number of combinations in each frequency group.

1. Group 1: C(5, n1) where n1 is number of winners from the *hot* group (e.g., C[5, 1] = 5; C[5, 2] = 10; C[5, 0] = 1).
2. Group 2: C(14, n2) where n2 is number of winners from the *mild* group (e.g., C[14, 1] = 14; C[14, 2] = 91; C[14, 0] = 1).
3. Group 3: C(20, n3) where n3 is number of winners from the *cold* group (e.g., C[20, 1] = 20; C[20, 2] = 190; C[20, 0] = 1).

Total of combinations is the result of the multiplication of number of combinations in each group. For example, the case 1-2-2 gives 5 * 91 * 190 = 86,450 total combinations. The 0-0-5 case generates C(20, 5) = 15,504 lotto combinations.

Another important parameter is the *range of analysis* (or *parpaluck*). Let's say, I chose *N * 2* (78 past draws). The degree of certainty for N = 78 is DC = 99.99%. That means that 99.99% of the lotto 5/39 numbers will come out in 78 drawings. I would try lower *parpalucks* like N (39 drawings) or N/2 (20). The parpaluck is higher for pairs because the probability is lower.

Calculating the number of combinations should be the easiest part. I said previously "to apply the combination formula." It is hard to apply a formula manually. There are handheld calculators with a function that calculates the combinations. My probability software also calculates the number of combinations with ease. The easiest-to-use program is *OddsCalc.EXE*. It can be downloaded from my downloads site, software category 5.6, *Scientific software: Mathematics, statistics, probability, combinatorics, odds, algorithms*.

The best distribution for pick-3 or pick-4 lottery games should be: *2-3-5* (*2 hot*-frequency digits, *3* from the *mild* group, and *5 cold* digits).

I received help from members of my *lottery and gambling message board*. They calculated total combinations for various frequency cases. I compiled the calculations in tables. They cover pick-3 and pick-4 lotteries, lotto 5/39, and 6/49 lotto games. This is the page of the very instructive tables: *http://saliu.com/frequency-tables.html.*

Another member of my online communities is also a very skilled computer programmer. He went a step further and

published the source code of several programs that work like my software. They generate the reports for the frequency groups and also generate combinations based on any type of frequency distribution. You might find of interest these two threads:
http://lotterygambling.phpbbnow.com/viewtopic.php?t=118 and *http://lotterygambling.phpbbnow.com/viewtopic.php?t=146*.

I created very complex software that generates the reports based on the three frequency groups: *hot*, *mild*, and *cold*. Not only that, but the software also *generates* the combinations based on the user input. For example, a lotto game that draws 5 from 39; the user chose as frequency strategy this distribution from the three frequency groups: *2-2-1*; the software easily generates all 18,200 lotto 5/39 combinations.

The programs in this software category are named in the fashion *SkipDecaFreq.EXE* (e.g., *SkipDecaFreq6.EXE* for 6-number lotto games). This type of software works with two more parameters:

1. The *Skips* (we already talked a lot about them).
2. The *Decades*—the lotto numbers are divided into decades (e.g., 1 to 9, 10 to 19, 20 to 29, etc.); a lotto drawing is a mix of numbers from various *decades*.

We are getting closer now to the concept of *filtering* in lottery—the very foundation in my lottery software. We saw a lotto drawing expressed as a string of three numbers representing frequency groups. For example, *2-2-2*. The first part of my software looks at all lotto drawings in the span of analysis. Every lotto drawing is expressed as 2, 2, 2; 3, 2, 1; 2, 4, 0; and so on.

The user decides that a string such as 3-1-2 will hit soon. The user runs the second part of the software: the combination-generating module. The *3-1-2* plays a *filtering* role: it eliminates all combinations that do not have 3 hot numbers *and* 1 mild number *and* 2 cold numbers.

The same thing is true for the *lotto decades*. A lotto 6/49 combination expressed as 4-9-12-13-19-39 is a drawing consisting of lotto numbers from the following decades:

= 2 numbers from decade 0–9
= 3 numbers from decade 10–19
= 0 numbers from decade 20–29
= 1 numbers from decade 30–39
= 0 numbers from decade 40–49

Based on the reports, the user decides that a *decade filter* such as 3, 0, 1, 0, 2 will hit soon. The generator will make sure that only combinations with lotto numbers from the corresponding decades come out.

The same thing is true for the *lotto skips*. A lotto 6/49 combination expressed as 4-9-12-13-19-39 is a drawing consisting of the *skips* 1, 1, 2, 3, 3, 4.

This means that the first lotto number in the drawing *skipped* one draw; that is, it came out also one drawing back or it waited one draw and hit again. And all that regardless of the position.

There is a second region of the skip filters that counts the *skips* for the same position only. As in the case above, the first number in the combination (it was 4 in the 4-9-12-13-19-39 case) came in same first position 7 drawings ago; thus, the skip for position 1 is 7.

Based on the reports, the user decides that a *skip filter* such as 1, 1, 2, 3, 3, 4 will hit soon. The generator will make sure that only combinations with lotto numbers with the corresponding skips come out.

We can apply the three categories of filters *simultaneously*. Or we can combine the filters two at a time. You can see a real-life case in Pennsylvania lottery (for a 5/39 lotto game). See *http://saliu.com/decades.html.*

Only two filters were selected: decades and frequency groups. These are the real-life parameters from the actual report:

Decade_1: 1
Decade_2: 4
Decade_3: 0
Decade_4: 0
Decade_5: 0
Decade_6: 0

Frequency_1: 0
Frequency_2: 0
Frequency_3: 5

Drawings in Pennsylvania 5/39 lotto satisfying the strategy:

#178: 4 11 13 17 18
#418: 6 10 11 15 17
#727: 8 10 12 15 19
#799: 7 12 13 14 17
#841: 5 11 13 14 15
#888: 6 11 12 13 14

The strategy hit 178, 418, 727, 799, 841, and 888 drawings back. What if I had played the respective strategy based on lotto decade and group frequencies? Total combinations for winning situations would have totaled 20-5-25-75-5-30. Average number of lotto combinations to play is 26.

The strategy hit 6 times in 1,000 drawings (less than three full years). You can see that the strategy once required just 5 combinations when it was successful. That particular lotto

5/39 strategy generated 75 combinations to play on one occasion. But I can tell you that on several occasions, the strategy does not generate one single lotto combination! The average number of combinations to play (tickets) is 26.

Playing in all 1,000 lotto drawings makes no sense at all (look at the skips between winning situations). For the sake of the argument, let's suppose we do so. The average *cost-of-winning* would reach $26,000 for $1 per ticket. The jackpot prize is listed at $100,000. The jackpot reaches over $250,000 on occasions, but also $50,000. Let's stick with the listed jackpot of $100,000. Winning 6 times amounts to $250,000 for a cost of $26,000. The return ratio reaches over 23. It's extremely rare to see such a return ratio in traditional investing: 2,600% over three years' time.

Chapter XV

LOTTERY SOFTWARE, LOTTERY FILTERING

1. Lottery Strategies Based on True Filters (Saliusian Filters)

My lottery software and filtering are inseparable. I started to write software in order to filter. I discover filters in order to incorporate them in my software.

I name my filters *Saliusian*—in accordance to my ego. After all, I discovered them. The filters are my brainchildren. I define also my filters as being *dynamic*. By contrast, other lottery developers use filters that are *static*. For example, *even/odd* and *high/low* are static; they appear a lot in software developed by others.

The concept of filter was unknown before I released my software. I came up with the concept of *filter* one morning, in the winter. I was a farm laborer as you know. I enjoyed when we didn't work, sometimes due to weather conditions. I was able to take care of my intense activity of software writing. That morning, however, I was out of coffee filters! I had coffee luckily. (I've never written a single line of code before having my coffee!) I solved the problem by using napkins or paper towel as coffee filter substitutes. It worked, somehow.

I also applied the *Turkish coffee* method (boil first and then wait to clear!)

I understood that morning how important filtering was! I used the new concept: *lottery filters*. Before that, I had *restrictions* in my software. I named the restrictions *eliminating conditions* sometimes. The vast majority of my lottery filters (*dynamic lottery filters*) are still unknown to all other lottery developers. I wish I had the time to explain to you all my lottery filters! It is mission impossible! Keep in mind that I have changed the filters a lot. I also eliminate some filters while add new ones to my software! My intention is to patent most of my software. The filters will be there in the source code. Once again, the more comprehensive source in this matter is SALIU.COM.

There is an enormous amount of filters present in my lottery software. In addition, every lottery data file is divided into *layers*—strata of data. A layer consists of 100,000 drawings sometimes (depends on the type of game). The *drawings* should not be taken literally. No lottery game has ever had, or will ever have, hundreds of thousands of real results! My software, however, allows for adding *simulated* (randomly generated) combinations (or numeric sets) to the real results.

Again, the *filters* are simply parameters that *eliminate* combinations in the generating processes (performed by my lottery software).

Ironically, mostly the *minimum levels* of the Saliusian filters are perceived as filters! Actually, the filters in my lotto software had only minimum levels for years. That's how it started. It took me many years to realize that there is no minimum without maximum.

The *minimum level* allows only lottery combinations *above* the level.

The *maximum level* allows only lottery combinations *below* the level.

Clearly, the two levels work *in opposition*. Like in good governance, the two opposing parties pursue the best for their constituents. Both parties (levels), however, are equally effective.

Filtering is founded on mathematics. I take as an example a filter that eliminates all *2-number groups* from past lotto drawings. That filter named *Two* (in most programs) works this way. In a 6/49 lotto, one 6-number combination can expand to C(6,2) = 15 two-number groups (*combinations of 6 taken 2 at a time)*. Each group can be combined with the remaining 49 – 6 = 43 numbers in groups of 4 numbers. That's C(43,4) = 123,410 lotto combinations. They combine in 43 * 123,410 = 1,851,150 6-number lotto combinations.

If we set *minimum_Two = 1*, then the program *eliminates 1,851,150* combinations. If we set *maximum_Two = 1*, then the lottery program only allows *1,851,150 combinations to be generated*. In this case, the advantage goes to the maximum level of the Two filter. In more than half the situations, the player has to deal with only 13% of total number of combinations.

In my experience, only the filter named *One* takes such values (median = 0). The One efficiency can be calculated as C(6,1) * C(43,5) = 5,775,588 combinations. Max_One = 1 is less effective than Max_Two = 1. But Min_One = 1 is more effective than Min_Two = 1.

Filters such as Four, FivS, and FivR have a special behavior. They are so high that they reach levels beyond the data file. If you analyze 10,000 draws, FivS in layer 1 shows 10,000. Most likely, the filter is higher than that. It can reach over

20,000—I know it for a fact. In the old Pennsylvania "6 from 69" lotto game, both FivS and FivR can reach 100,000 easily. That's why the FivS and FivR filters do NOT have maximum levels! The Four filter has a maximum level, but it requires caution. If the filter shows 10,000 in your WS6.1 report, it simply means there weren't enough draws to analyze.

There is one filter in my software named *Ion5*. It triggered a gold rush a few years back. Many users viewed the Ion5 filter as the key to a gold mine. They saw many filter values equal to *417*. They thought immediately of setting *Ion5* as follows:
~ minimum level = 417
~ maximum level = 418

They ran the program and generated *no* combination at all after days and nights of continuous running! There was no bug in *MDIEditor and Lotto WE* or in my *command prompt lottery software*. A repeating value of 417 for Ion5 indicated an insufficient size of the data file. The 417 value was not to be relied on as far as a maximum level was concerned. Most likely, Ion5 goes a lot higher.

The PCs were significantly slower just five years ago. My software could not handle effectively huge data files that included simulated combinations as well. Things are far better now performance wise. My software can run now lotto data files with millions of lines! I can only recommend to the users of my software not to use *Max_Ion5=418* under the old circumstances! That's why their computers didn't generate any lotto combination. It is very, very rare for a value of *Ion5* to reach *exactly 417*. The maximum values should be used only with very large data files. Any time you see a value higher than 100 (e.g., 417 or 1,000) repeating more than a dozen times, it should raise a red flag. The D6 data file is too small (real draws + simulated combinations). Create a very large simulated data file of at least 500,000 (five hundred thousand) combinations (lines). You can now download a simulated data file with all 13,983,816 combinations in the 6-from-49 lotto game. My lottery software works effectively with such huge lottery data files.

Right now, we can define a lottery strategy in a new format. *A lottery strategy is a collection of settings for one or more filters.* There are possibly as many strategies for playing lotto and other lottery games as there are combinations (sets) in them, but one thing is for sure. If you find a decent strategy for the game you are playing, you increase your chance to hit the jackpot massively!

A lottery strategy can be *tight* or *loose*.

1. A *tight* strategy will generate very few lottery combinations or none. Such strategy will hit rarely, but with a very low cost. A *tight* strategy requires patience. Such strategy is the result of *setting a large number of lottery filters*. Too many filters can generate no lottery combinations at all. Also, a *tight* strategy can be the result of setting just one lottery filter, but with very close minimum and maximum levels. You might set the *minimum level of Ver_6* to 5 and the *maximum level of Ver_6* to 6. It might never generate one single combination.

2. A *loose* lottery strategy is the exact opposite. It will generate lots and lots of lotto combinations. The most common form of *loose* lottery strategy is *disabling* all the filters. In other words, none of the filters in software is set (enabled)—neither the minimum level nor the maximum level.

• Here is a *statistical* method of creating a lottery strategy: *set one or very few filters to levels outside their normal ranges.* The normal range is determined by the *median* of the respective lottery software filter. The median is automatically calculated by the *winning report generators* incorporated in my lottery software. You can see it at the top of each filter (a column in the *winning report* files). Filters *outside the normal ranges* could be

~ median multiplied by 3 or 4 or
~ median divided by 3 or 4.

For example, if a median is 12, you can set a tight lottery filter to 12 x 4 = 48 (or rounded up to 50) for the minimum value of the filter. Or you can set an equally tight filter to 12 / 4 = 3 + 1 for the maximum level of the respective filter. (Keep in mind, the maximum level of a filter must be at least [minimum level] + 1). If a filter is set to 4 times the median, it slashes in half 4 times the total of combinations generated. In the pick-3 lottery example, 1,000 lottery combinations are reduced to 500 in the first step, 500 slashed to 250, 250 halved to 125, and finally, 125 reduced to 60 plus.

There are situations when a filter reaches levels of 10 times or more the value of its median (or the median divided by 10). Playing such rare but opportunistic situations results in very few combinations to play; there will be *no* lottery combinations (sets) on many occasions.

A sorted-by-column winning-report file can show you even more valuable information. Say you sorted W3.1 by the *Pairs-1* column (filter). The median was 32. The median divided by 4 is equal to 8. Go to line 1 of the column and see how many Pairs-1 are lower than 8. You can see also what kind of levels other filters show for Pairs-1 less than 8. Other filters may show very low numbers as well. Other lotto filters may show bigger numbers. You can choose as a playing strategy *Max_Pair_1*=8+1=9, plus other filters at less-tight levels. For example, Max_Vr_1=4, Max_TV_1=6, Val_1=5. This is just an example. You can find similar numbers in your sorted WS files.

The median multiplied by 4 is equal to 128. Go to the last line of the column and see how many *Pairs-1* are larger than 128 (or 120 or 130; you can round up or down for more flexibility in your choices). You can see also what kind of levels other filters show for Pairs-1 greater than 128. Other filters may show very high numbers as well. Other filters may show lower numbers. You can choose as a playing strategy Min_Pair_1=130, plus other filters at less-tight levels. For example, Min_Vr_1=1, Min _TV_1=5, Min_Syn_1=50.

Using such tight levels for one or very few lottery filters eliminates a huge amount of lotto combinations. Such levels occur more rarely. You should not play them in every drawing. They skip a number of drawings between hits. My lottery software makes it even easier for you. Every package has also a *strategy-checking utility*. It shows how a particular strategy fared in the past. The *strategy-checking* modules show the *levels* of all the filters and the *skips* of the strategy.

You should only play a strategy if its current skip (the first number in the skip chart) is less than or equal to the median. For example, if the median of the strategy is 5, you should play the strategy only if the first number in the string of skips is 0 or 1 or 2 or 3 or 4 or 5. If the current skip is larger, don't play the strategy; save the money. Since you can select a very, very large number of lottery strategies, look for another strategy to play. Look for one or more strategies that show current skips under the medians. Those are the best strategies to actually play for the time being.

• There is yet another path to creating lottery strategies. The method takes into consideration a *higher probability of a filter level to occur*. Look at the lottery filters, from line 1 back to previous drawings. It is evident that the filters go up and down. It is a law. The ups and downs are expressed by the + (plus sign) or – (minus sign), respectively. When a filter is higher than in the previous drawing, the filter has the + sign at the right. If the filter is lower than in the previous draw, it has a – sign attached. It is more visible. You can notice that in most cases, the filters go from one trend to the opposite after two or three drawings. That is, after two or three + signs, the – sign comes up or vice versa. Based on that, we can look at each filter (column) in the WS files.

The key position is line no. 1. If the sign in line no. 1 is – and also in line no. 2 and line no. 3 (3 decreases in a row), we should expect a + (increase) the very next draw. If the sign in line no. 1 is + and also in line no. 2 and line no. 3 (3 increases in a row), we should expect a – (decrease) the very next draw. Let's take pick 3 as an example. If Pair-1 in line no. 1 is 12 and it shows –, the third consecutive – (decrease), we should expect a + in the very next drawing.

An *increase* in a filter requires the use of the *minimum* level of the respective filter. In this example, I'll set Min_Pair_1=13. If I want to increase the probability, I can set Min_Pair_1=10 for example.

Let's say now Pair-1 in line no. 1 is 123, and it shows +, the third consecutive +. We should expect a – in the very next drawing. A *decrease* in a lottery filter requires the use of the *maximum* level of the respective filter. In this example, I'll set Max_Pair_1=124. If I want to increase the probability, I can set Max_Pair_1=130 for example.

You can look for longer streaks of either + or –. Just go the line no. 1 in each WS file. There are situations when the current streak can be 4 long or 5 long, even longer in rare situations. You may want to consider first the longer like-sign streaks. Keep in mind, however, that the streaks shift direction after up to 3 drawings in most cases. Actually, streaks of 1 or 2 consecutive like signs are the most frequent.

You can combine filters selected as in this path with the type of selection presented in path no. 1. You can set one tight filter (4 times the median, etc.). Then you set other filters as in path no. 2. For example, Min_Pair_1=120 (path no. 1), Max_Vr_1=7 (path no. 2), Min_TV_1=10 (path no. 1), Min_Syn_1=100 (path no. 1), Max_Bun_2=6 (path no. 2), Max_Tot_3=1500 (path no. 2), Max_Any_5=300 (path no. 2), and so on.

2. How to Run My Lottery Software

My lottery software is a large collection of programs I started to write in 1985. I started to publish some lotto programs beginning 1988. My software covers virtually every lotto and lottery game played in the world. Of course, some game formats are better represented than others. As George Orwell put it so eloquently. *"Some animals are more equal than others."*

Running my lottery software has been the main challenge faced by the users. I have always been a lot more determined to discover or create something than to write about it in minute details. Writing my software consumed a lot of my time. Writing in detail about the usage of my software has always seemed a luxury to me. As things were quite clear on my mind, I expected all others to have the same clear picture of my software! Of course, it is not the right mentality—and I never thought in those terms consciously and seriously. I was short on time. Additionally, my English did not help me a great deal for many years. If I had good mastery of the English linguistic tool, I would have already written a collection of books by now. I am just getting started.

I'll do my best right now to make a good introduction of my software—how to use it. As diverse as my lottery software is, there are *common steps* that every user should take. The presentation that follows is concise. Throughout this book, I suggested reading specialized pages at SALIU.COM. It is the case again. There are tips on using my lottery software all over my Web site. I mention here the following Web pages:

http://www.saliu.com/MDI-lotto-guide.html
http://www.saliu.com/gambling-lottery-lotto/command-prompt.htm
http://www.saliu.com/filters.html
http://www.saliu.com/bbs/messages/818.html
http://www.saliu.com/forum/lotto-book.html
http://www.saliu.com/infodown.html
http://www.saliu.com/Help.htm

These are the *six essential steps* common to all my lottery software packages:

1. Create and update the lottery data files.
2. Generate the winning reports (the WS files).
3. Create lottery strategies by analyzing the WS reports.
4. Check a lottery strategy to establish its performance in the past.
5. Check the strategy hits in the past to establish an average of the amount of tickets to play.
6. Run the combination generating modules.

A concise presentation of each procedure follows.

1. Create and update the lottery data files.

You cannot do anything with *MDIEditor and Lotto WE* or *Bright.exe* or any *Lotwon* lottery software without ***data files***! Such files hold a gold mine: the *drawings* or *draws* or *results* or *past winning numbers* in your lotto and lottery games! You only need the *pure numbers* of lottery drawings. That is, only the winning numbers—*no* dates, *no* prizes, *no* bonus numbers (as in some lotto games). Absolutely nothing else!

Let's exemplify by using the simplest of the lottery number games: pick 3 (drawing 3 digits, each from 0 to 9).

• Start by ***creating*** a pick-3 data file. It is the easiest possible step. Open a text editor, including Notepad. Type 3 digits per line, exactly 3 digits. Separate the digits by a comma or by a space (at least one blank space). Press Enter at the end of the line.

Continue typing the next three digits, which in fact represent the previous pick-3 lottery drawing. Make sure there are no blank lines in the file—none whatsoever. When done, save your file always in text format (no formatting whatsoever, just pure numbers).

You can give your pick-3 lottery data file a name like *DATA-3*. Actually, any name would do; just remember it when you need to open the file! Again, my lottery software requires that the data (results) files for the pick-3 lottery game have exactly 3 digits on each line; the lottery digits are separated by blank space(s) or commas.

The format I strongly recommend: separate the elements by blank spaces. If your data files are not in proper format, my lotto/lottery software will not run or will yield erroneous results. These are the correct data file formats to follow:

Pick 3: exactly 3 digits per line (0 0 3)
Pick 4: exactly 4 digits per line (0 9 3 4)
Lotto 5: exactly 5 numbers per line (1 12 13 24 35)
Lotto 6: exactly 6 numbers per line (1 12 13 24 35 46)
Lotto 7: exactly 7 numbers per line (1 2 13 24 25 36 37)
Keno: exactly 22 numbers per line (1 to 80; the last 2 numbers must be 0 0 if the game format is 80/20)

Powerball, Thunderball, Mega Millions: exactly 5 regular numbers per line plus 1 power ball (1 12 13 24 35, 1)
Horse racing trifectas: exactly 3 horse numbers per line: (13, 1, 6)
Horse racing superfectas: exactly 4 horses (numbers) per line (13, 1, 6, 4)

It's highly recommended to run my lottery/gambling freeware *Sorting.EXE* to sort the combinations (lines) in the data files in ascending order. The utility also nicely formats the files (with evenly spaced fields).

• When ***updating*** a lottery data file, keep in mind this rule: the latest (the most recent) drawing always goes to the top of the file, becoming the line number 1. The oldest drawing is the bottom line in the file. If your data file is not in this order, you need to use the included program *Updown.EXE*. It reverses the order in a text file: the bottom becomes the top of the file.

Tip: The Bright.exe software packages come with another useful utility: *Parsel.EXE*. The utility checks lottery data files for correctness. It can find a number of errors in your data files, and it points you to the lines with errors. You should run *Parsel.EXE* from time to time to make sure your lottery data files are error free.

The lottery programs in the new *Bright.EXE* software packages require final data files of hundreds of thousands of lines. Of course, you don't have that many real drawings in a lottery game! So you create an additional file of randomly generated lottery combinations. The additional file is usually named SIM* and is created in seconds by the "Utilities" functions. The Utilities function also creates the D* file: the final file used by the WS report generators and the lottery combination generators. The new, much more powerful Super Utilities (available via B3.EXE), also create SIM* and D* files.

2. Generate the winning reports (the WS files).

You need all those lottery drawings in the data files you created/updated at step 1 for one special purpose: winning. This step follows naturally. You need to *generate the winning reports*. When I started, I named the reports the *WS* for *Winning Strings*. My lotto software didn't apply many filters in the beginning. The reports only showed short strings of numbers. Believe it or not, before that, I did the entire process manually! It was painstaking, but I didn't know how to program that reporting for a while! The new reports are saved now as W and MD (e.g., W6 and MD6 in a 6-number lotto game).

In general, this function is named *Generate Winning Reports* in my lottery software. It starts by pressing the *F5* function key or by pressing *W* (under Vista/Windows 7).

At the top of every report, you have *column headings* with their respective *filter names*. Under each is given the filter's respective median, average, and standard deviation values. At the very left of the filter table (W or MD), you will see the heading *Draw*. And under this heading are the numbers 1–100. They correspond to the Data-6 file (the real results in your game). That is, draw number 1 corresponds to line 1 in the Data-6 file. Reading across the table from *Draw 1*, you see a set of values for each filter. These filter values are produced as a result of the numbers that came out in line 1 of the Data-6 file and how they relate to previous draws.

You look at all those numbers with + / – next to them. You do that with the purpose of *setting filters* and choosing *playing strategies*.

The W and MD report files are in simple text format. You open them in any text editor in order to view the reports. Even *Notepad* will do. My software comes with one good 16-bit text editor. It doesn't work with the 64-bit Vista/Windows 7. But you have also *MDIEditor and Lotto WE*, which works fine under the 64-bit Vista/Windows 7.

3. Create lottery strategies by analyzing the WS reports.

This step was analyzed in more detail in section 1 of this chapter. *This is the most important step. Selecting just one good strategy can make you a millionaire quickly.*

Please review that section and also read the additional materials at SALIU.COM.

4. Check a lottery strategy to establish its performance in the past.

This step is not mandatory, unlike the previous three. This function incorporated in my software makes it much easier to look at your possible strategy. You can clearly compare the instances where your strategy hit and look at how all the other filter values compare with each other. It is a good idea to create a blank form on a piece of paper, which resembles the strategy checking filter input screen, so you can enter various Min and Max values for the filters on paper first as you refine your strategy. This is also useful when you find a strategy that you want to play as you will then have a hard copy of the filters' values that you use.

The latest *Bright.EXE* software packages automatically create the correct format of the strategy checking files. You can experiment very easily with a large variety of filter settings. You can go back and forth between changing strategies and viewing the respective reports.

Usually, this function is named *Strategy*, *Check*, or *Check Strategy* or something similar (intuitive!)

5. Check the strategy hits in the past to establish an average of the amount of tickets to play.

This step is not mandatory either. But as step 4, the built-in function makes your life easier. So you decided to apply a strategy. You checked its performance in the past, and you saw it had hits with a good frequency. You want to know how many combinations (tickets) the strategy would require.

This function is usually named *Strategy Hits*. It refers to the *hits* (wins) in the past. The function also calculates the *average* amount of combinations required to play.

This function is even more valuable. You can see *how many combinations to expect*. You can postpone playing with this simple trick. You run the combination generator with your up-to-the-date data file. Input the strategy file you wanted to play. If the program generates far more combinations compared to the figures shown by the *Strategy Hits* function, it is not a good time to play the strategy the very next drawing! Of course, if the program generates no combinations at all, you wouldn't have anything to play anyway!

6. Run the combination generating modules.

And we reach the *reward* stage! This is where we should materialize our efforts—win consistently lottery prizes.

You can generate combinations even without checking for strategies first. You just type the filter values at the screen prompts. The function keys (usually F6 and F7) start the combination generators. The first one generates combinations in *lexicographical order*. The F7 key generates lottery combinations in *optimized random order*.

The combination generators have a multitude of functions:
~ purge lottery combinations from an output file (previously generated lotto combinations)
~ generate combinations based on favorite lottery numbers
~ the user can enable or disable the *inner filters*

Once again, nothing works if you don't have a valid data file available. If you do have a valid data file (correct format, right size, or sufficient numbers of lines), you may work by simply typing filter settings at the screen prompts. It is recommended to select the screen as the output device if you are practicing only.

The *purge* function in the combination generators is a tool of *expandability* and *interoperability*. The function takes an output file of previously generated combinations and *purges* it. That is, it applies the filtering to reduce further the amount of combinations in the output file.

The output file can be
~ a file of combinations previously generated by *MDIEditor and Lotto WE*,
~ a file of combinations previously generated by the *command prompt* lotto/lottery programs,
~ a lotto wheel (that needs to be reduced probably because of its large size), or
~ a file of combinations previously generated by another application (other than *MDIEditor and Lotto WE* or my *command prompt* lottery software—could be another vendor's software).

The output file must be compliant, however, with the format of my lottery software:

~ The file must be in text format (no special formatting, no blank lines).

~ The file must be compliant with the format of the data file for the respective lotto/lottery game (e.g., an output file to purge in the lotto-6 game must have exactly 6 numbers per line [combination]; exception is Keno: DataK must have exactly 22 numbers per line [the drawings], but an output file has exactly 10 numbers per combination).

The *Purge* function is also a valuable tool to test the effectiveness of the filters. Instead of generating millions of combinations, generate a "purge" file of say 10,000 combinations in a lotto game. Then you can test how many combinations a particular filter eliminates. You can test the internal filters as well. Check the corresponding check box to enable the inner filters (in *MDIEditor and Lotto WE)*. Make sure the rest of the filters are disabled (i.e., the text boxes in the input form are blank or set to zero). Click OK to start the purge process. The program will eliminate the combinations restricted by the inner filters. You can divide the amount of the combinations generated to the total number of combinations in the purge (output) file and thus calculate the percentage of the elimination.

You can also test the efficacy (eliminating power) of each filter for a value equal to 1 for both the minimum and the maximum levels. Let's exemplify by the famous-by-now *Ion5* filter.

First, here is how to figure out the maximum level to set Ion5 to. Say the game is lotto 6/49. The integer part of 49 divided by 2 is 24. Divide the total number of combinations to use (what you entered at the respective prompt) by 24. That is the maximum level for Ion5 you can set to. Keep in mind that Ion5 can be bigger than that, sometimes much larger. It all depends on the total number of drawings to use. For a lotto 6/49 game, it is likely that 200,000 past drawings to use will report the correct maximum value for Ion5.

Test the efficiency of the minimum level of Ion5: purge the 10,000 output file against the D6 file. Set the min_Ion5 text box to 1 in the input form. Make sure all the other text boxes are blank and "Enable inner filters" is not checked. Click OK to start generating combinations. Compare the amount generated to the initial 10,000 combos.

Test the efficiency of the MAXIMUM level of Ion5: purge the 10,000 output file against the D6 file. Set the MAX_Ion5 text box to 1 in the input form. Make sure all the other text boxes are blank and "Enable inner filters" is not checked. Click OK to start generating combinations. Compare the amount generated to the initial 10,000 combos.

The eliminating efficiency percentages tend to a constant for larger and larger output (purge) files. I believe 10,000 is an adequate benchmark.

The most accurate testing method, however, is the *Sequential* option. It generates all the combinations in the game in *lexicographical* order. Here is how you can also check if a strategy generates, correctly, the winning combination.

This is an example for the lotto-4 game. Let's say that the filters report shows the following string of filter values for line (draw) no. 20, data file DataL4, from Ion1 to Ver4:

112-2-3-6-11-1-9-13-15-1-9-15-43

First, open the DataL4 file and delete the first 20 lines (drawings). Draw no. 20 will act now like the unknown: the very first winning combination to hit next. We "do not know" what combination it will hit next! Save as the new file under a different name (e.g., DataL4.2). Run Lotto-4, Sequential. Data file to run: DataL4.2. In the input form, type all the filters in the minimum side of the input boxes (from 112 to 43). Be absolutely accurate! Next, type in the maximum side of the input boxes the filter values increased by 1:

113-3-4-7-12-2-10-14-16-2-10-16-44

This is the tightest filter setting if the strategy was correctly entered—and if the software is error free! The program will generate exactly 1 combination; it is the winning combination—the one showed in line no. 20 in the filters report!

The inner filters must be disabled; the Past Draws filter must be set to 0 (empty input box). Also, be aware that some filters can reach very high values. If your data file is too small, some filters (like Ion1, Ion4, or Ion5) could have reached far higher levels. Do not enable their maximum levels! You will notice that a largest number in a column repeats very often.

This method (running Sequential with the tightest settings of the filters) discovered the largest number of bugs in *MDIEditor and Lotto WE*! Many people have guaranteed to me there are no more errors in this great piece of software!

FileLines.EXE ("Check," "Strategies," "Cross-check") is an additional tool in the process of expandability and interoperability of *MDIEditor and Lotto WE.*

3. My Lottery Software: Major Titles

My lottery software is a large collection of programs, isn't it? I believe I wrote over 500 software titles, most of them lottery programs. I reduced down the offering to just over 100 programs. I will mention here only the most important lottery software packages. The newest creations are always the best. We always evolve!

• *MDIEditor and Lotto WE* is the most comprehensive lottery and lotto statistical analyzer and combination generator for Windows 95/98/Me/NT4/2000/XP/Vista. The multigame software works also as a potent text editor, mainly to contain, format, and print the lottery combinations.

The super application, founded on mathematics, covers these types of lotto and lottery games: pick-3 and pick-4 lotteries, horse racing (trifectas and superfectas), lotto-4, lotto-5, lotto-6, lotto-7, Powerball/Mega Millions "5+1," EuroMillions "5+2," Keno.

• *Bright3.EXE* is, indeed, high-powered integrated pick-3 lottery software. This software bundle is far more powerful even than Pick332.EXE. It's OK to keep older packages, but this is the brilliant way to go as far as pick-3 is concerned.

• *Bright4.EXE* is, indeed, high-powered integrated pick-4 lottery software. This software package is far more powerful even than Pick432.EXE. It's OK to keep older bundles, but this is the brilliant way to go as far as pick-4 is concerned.

• *Bright5.EXE* is, indeed, high-powered integrated lotto-5 software. This software bundle is far more powerful even than Pick532.EXE. It's OK to keep older packages, but this is the million-dollar way to go as far as lotto 5 is concerned.

• *Bright6.EXE* is, indeed, high-powered integrated lotto-6 software. This software suite is far more powerful even than Pick632.EXE. It's OK to keep older bundles, but this is the million-dollar way to go as far as lotto 6 is concerned.

• *BrightH3.EXE* is, indeed, high-powered integrated horse-racing trifecta software. This software package is far more powerful even than PickH32.EXE.

• *LOTWON7.EXE* is the super lotto-7 software package, with useful info that can be applied to the lotto-5 and lotto-6 packages. Read the new PICK7.TUT and README.TXT. It features an improved menu based on the pick-3 package (LotWon3).

• *LOTWON99.ZIP* is SuperPower_95 version 10.99.02. This is a most powerful collection of lotto and lottery software utilities. The package tackles thorough (and very useful) statistics on one hand. On the other hand, the application is one of the best combination generators in the business. At least, these are the very best random number generators of all time! You may want to always check the date of my freeware software. If it's newer, it must be better—much better sometimes! You might want to download this package just to experience my thinking in probability theory applied to lottery. Please be sure to read the text file TUTORIAL.DOC. The newer 32-bit lotto/lottery software packages are much, much more powerful with many, many more significant features.

• *Combinations.exe* is a lotto software that generates combinations for absolutely any type of lotto game, plus horse-racing straight sets. Specific to this program, the combinations can be generated in steps. That is, the user has the choice to generate lotto combinations with constant gaps or skips between them. For example, starting at the very top of a combination set (the lexicographical order no. 1), then step 90, the following combination generated will have lexicographic order (rank) no. 91, . . . and so on to the very last combo in the set. To my best knowledge, no other piece of lotto software can accurately generate universal *N* taken *M* at a time combinations. Most certainly, no other program can generate lotto combinations in steps. Furthermore, this incredible program even generates lotto combinations within a range of numbers!

• *Lotto2Games51.exe* is a special utility software for two-in-one (5+1) lotto games: Powerball, Mega Millions, CA SuperLotto, Thunderball, etc. The lotto software works with single and multiple number groups: pairs (doubles), triples, quadruples, quintuples. The user can generate "5+1" lottery combinations with or without favorite numbers (from 0 or no favorites to 4 favorites).

• *SoftwareLotto7.exe* is a special lotto software for 7-number lottery games worldwide. The lotto software program works with single and multiple number groups for lotto-7: pairs (doubles), triples, quadruples, quintuples, sextets. The user can generate lotto combinations with or without favorite numbers (from 0 or no favorites to 6 favorites).

• *SoftwareLotto5.exe and SoftwareLotto6.exe* are special lotto utility software for 5—and 6-number lottery games worldwide. The two lotto software programs work with single and multiple number groups for lotto-5 and lotto-6: pairs (doubles), triples, quadruples, quintuples. The user can generate lotto 5–6 combinations with or without favorite numbers (from 0 or no favorites to 4–5 favorites).

For this and everything else available at my software download site, I invite you to visit these places:

• *http://www.saliu.com/infodown.html*
~ The gateway to the downloading area of my software.

• *http://www.saliu.com/free-lotto-lottery.html*
~ Free lotto/lottery software like Lotto 5, 6, 7, Pick 3 and 4 Lotteries, Powerball, Mega Millions, Thunderball, EuroMillions, Keno.

• *http://www.saliu.com/free-lotto-tools.html*
~ Free lottery software: utilities, tools, lotto wheels.

There has been no doubt that my software has created winners of significant prizes in lottery and gambling as well. You may never hear or read the real names of the winners. Testifying on winning can be a real *legal challenge*. What if I had hard-nosed lawyers who would pursue prize sharing? *The tithing law, anyone?* Every winner is better off keeping quiet.

To all of you, dear readers and potential big-time winners, I wish you the best of luck!

Index

M

N

Y

Z